Statistical Analysis and Control of Dynamic Systems

Mathematics and Its Applications (*Japanese Series*)

Statistical Analysis and Control of Dynamic Systems

by

Hirotugu AKAIKE

Institute of Statistical Mathematics, Tokyo, Japan

and

Toichiro NAKAGAWA

System Sogo Kaihatsu Co. Ltd., Tokyo, Japan

Translated by Hirotugu Akaike and Maki Akaike Momma

KTK Scientific Publishers/Tokyo

Kluwer Academic Publishers

Dordrecht/Boston/London

Library of Congress Cataloging-in-Publication Data

Akaike, H. (Hirotugu), 1927–
   Statistical analysis and control of dynamic systems/
  H. Akaike and T. Nakagawa.
     p. cm.—(Mathematics and its applications. Japanese series)
    Includes index.
    ISBN 9027727864 (Kluwer)
    1. Control theory. 2. System analysis. 3. Cement kilns
—Mathematical models. I. Nakagawa, Tomoyasu. II. Title.
III. Series: Mathematics and its applications (Kluwer Academic
Publishers). Japanese series.
QA402.3.A44 1988
629.8′312—dc 19
                               88-19749
                                 CIP

Published by KTK Scientific Publishers (KTK),
302 Jiyugaoka-Komatsu Building, 24-17 Midorigaoka 2-chome, Meguro-ku,
Tokyo 152, Japan,
in co-publication with Kluwer Academic Publishers, Dordrecht, Holland

Sold and distributed in the U.S.A. and Canada
by Kluwer Academic Publishers,
101 Philip Drive, Assinippi Park, Norwell, MA 02061, U.S.A.
in Japan by KTK Scientific Publishers (KTK),
302 Jiyugaoka-Komatsu Building, 24-17 Midorigaoka 2-chome, Meguro-ku,
Tokyo 152, Japan

In all other countries, sold and distributed
by Kluwer Academic Publishers,
P.O. Box 322, 3300 AH Dordrecht, Holland

This book is partly subsidized by the Ministry of Education, Science and Culture of Japan.

Printed in Japan

SERIES EDITOR'S PREFACE

Growing specialization and diversification have brought a host of
monographs and textbooks on increasingly specialized topics. However,
the "tree" of knowledge of mathematics and related fields does not grow
only by putting forth new branches. It also happens, quite often in fact,
that branches which were thought to be completely disparate are
suddenly seen to be related.

Further, the kind and level of sophistication of mathematics applied in
various sciences has changed drastically in recent years: measure theory is
used (non-trivially) in regional and theoretical economics; algebraic
geometry interacts with physics; the Minkowsky lemma, coding theory
and the structure of water meet one another in packing and covering
theory; quantum fields, crystal defects and mathematical programming
profit from homotopy theory; Lie algebras are relevant to filtering; and
prediction and electrical engineering can use Stein spaces. And in
addition to this there are such new emerging subdisciplines as "experimental mathematics", "CFD", "completely integrable systems", "chaos,
synergetics and large-scale order", which are almost impossible to fit into
the existing classification schemes. They draw upon widely different
sections of mathematics. This programme, Mathematics and Its Applications, is devoted to new emerging (sub)disciplines and to such (new)
interrelations as exempla gratia:

- a central concept which plays an important role in several different
  mathematical and/or scientific specialized areas;
- new applications of the results and ideas from one area of scientific
  endeavour into another;
- influences which the results, problems and concepts of one field of
  enquiry have and have had on the development of another.

The Mathematics and Its Applications programme tries to make available a careful selection of books which fit the philosophy outlined above. With such books, which are stimulating rather than definitive, intriguing rather than encyclopaedic, we hope to contribute something towards better communication among the practitioners in diversified fields.

Because of the wealth of scholarly research being undertaken in the Soviet Union, Eastern Europe, and Japan, it was decided to devote special attention to the work emanating from these particular regions. Thus it was decided to start three regional series under the umbrella of the main MIA programme.

In practice, principally in "mathematics in engineering situations" it is astonishing how often the problems found in industrial settings centre around the identification (hence prediction) and the (optimal) control of systems with noise. That, at least, has been one rather striking experience of ours with ECMI (the European Consortium for Mathematics in Industry).

This area—noisy dynamical systems and their control—is one in which great progress has been made recently, both in terms of theory and algorithms and in terms of computer programs for the data processing. There is much that is ready to be applied; it just remains to do it. This is not, as the authors explain convincingly and at length, a trivial matter. The link from real data to the various results available is not so easy to forge even for very simple kinds of models. The would-be applier needs help, and, in addition, having knowledgeable guidance in navigating the jungle of available computer programs is highly desirable.

This book, whose senior author is certainly extremely well known in the field, provides all this and more, all in the framework of a very detailed case study, perhaps the best way to "learn" the art of applying mathematics to real problems.

The unreasonable effectiveness of mathematics in science ...	As long as algebra and geometry proceeded along separate paths, their advance was slow and their applications limited.
Eugene Wigner	But when these sciences joined company they drew from each other fresh vitality and thenceforward marched on at a rapid pace towards perfection.
Well, if you know of a better 'ole, go to it.	
Bruce Bairnsfather	Joseph Louis Lagrange.
What is now proved was once only imagined.	
William Blake	

Helsingöi, June 1988   Michiel Hazewinkel

PREFACE

In many scientific endeavors the final objective is the prediction or control of the behavior of the object that is being investigated. In this connection, the following two points are characteristic of the modern theory of control:

1) the use of the state space representation of the system in anticipation of the application of a computer,

and

2) the use of the statistical concept of noise to describe the input that drives the system.

In spite of the significant development of this new type of control theory, serious gaps may be seen between the theory and practice in many areas of its application. The comparison of this situation with the case of the application of classical theories, such as Newtonian mechanics which is based on basic concepts like the center of gravity, mass, acceleration, and force, makes the following point clear. In the application of modern theories which are based on statistical concepts, users must first establish the correspondence between the basic theoretical model and the real situation under study. The recognition of the importance of this point is still quite often lacking and the resulting confusion forms the barrier between theory and practice.

In this book we consider that the barrier can be eliminated only with the help of proper statistical analysis of observational data. Discussions will be developed through the study of an example of a real process where control is realized by first confirming the characteristics of the system through the analysis of its observational record and then proceeding to the implementation of a control. The model of the system to be treated in this book will be assumed to be constant and linear. Although this model may seem restrictive, its practical utility will nevertheless be demonstrated by an example of the implementation of the computer control of a cement kiln process.

Statistical dynamic systems that exhibit random fluctuations appear as important research topics in many areas, including science, engineering, economics, medicine, and biology. The present book is intended to provide initial assistance to researchers or engineers who are faced with real data in one of these wide areas of applications. The technical detail of the actual process of data handling is heavily dependent on the use of computers.

Accordingly, the explanation of the theoretical aspect of the data processing is kept to the minimum and every effort is concentrated on the explanation of the use of the computer programs. Almost all the basic computer programs for stationary time series analysis available at the time of writing the book are included in this volume.

The research on the computer control of the cement kiln process that provided the basis of the present book was originally started through the pioneering initiative of Mr. Tsuneo Ootomo, president of the Chichibu Cement Company, who was then a member of the board of directors. Mr. Ootomo directed the research and development relating to the project and provided the authors with the chance to undertake the cooperative work and encouraged the publication of the result. For all of this we are very grateful.

In connection with the research reported in this book the authors have benefited from the help of many people of both the Institute of Statistical Mathematic and the Chichibu Cement Company. In particular, Ms. Emiko Arahata helped us significantly with the programming of the programs included in Chapter 5. Messrs. Yoshitaka Yagihara and Takeshi Kominami worked out the implementation of the computer control. Mr. Yuzo Morihira of the Saiensu-shya Publishing Company encouraged the production of the book. Mr. Yoshihiro Aota undertook much editorial work. We are grateful to all of those who thus supported our activity directly or indirectly.

The purpose of the present book is to provide a general introduction to the statistical analysis and control of dynamic systems. Accordingly, the description of the cement kiln is limited to the conceptual level and the discussion of the technical details specific to the process is omitted. However we note that an on-line control system of the overall cement production process is now fully developed by the Chichibu Cement Company and further promising development seems likely in the future.

The present book is a product of close cooperation between the authors. However, to clarify the contribution of each author, the names of the actual authors of each chapter are included as follows:

Chapter 1	What is the problem?	by Akaike
Chapter 2	An explanation of the difficulties involved	by Nakagawa
Chapter 3	Statistical preliminaries	by Akaike
Chapter 4	A successful application	by Nakagawa and Akaike
Chapter 5	Computer programs	by Akaike

April 1972

The Authors

GENERAL EXPLANATION OF THE BOOK

The authors have been working together for the past several years to realize a computer-automated operation of cement kilns. As will be described in detail in the book, cement kilns showed very complex movements of the internal state variables, such as the temperature and pressure, and it was believed that only a human operator with long experience of operation could control the kiln in practice.

Would it be possible to operate such a complex system by computer? The conclusion eventually reached by the authors was that this was in fact possible. The motive that drove the authors to write this book was the desire to disseminate the information about how they came to this conclusion.

The cement kiln is a typical example of a system that shows complex inter-relationships in the behavior of mutually dependent variables over a certain period of time. This type of system appears as an important subject of study in various fields such as medicine, biology and economics, as well as physical science and engineering. The authors believe that the description of their experience of the analysis and control of the kiln will clarify many points that cannot be covered by a formal general theory and will thus help researchers working on similar subjects.

This book is not styled as an ordinary text book on a general theory. Its purpose is to explain, with a real example, how a statistical formulation of an actual problem is obtained, how the correspondence of the model with real data is realized, and how objective knowledge is obtained that cannot be gained by a simple accumulation of experiences. Although the book may provide only an incomplete picture of the whole process, yet it is hoped that it will be read as a record of an exciting exploration.

The problem treated in this book is representable by a basic model and the related procedures are also kept at basic level. The construction of the book is as follows:

Chapter 1 What is the problem?
The basic problem to be treated in this book is presented and some key points in handling the statistical problems are explained.

Chapter 2   An explanation of the difficulties involved

The difficulty of implementing the control of a stochastic system is explained with the cement kiln process as a real example.

Chapter 3   Statistical preliminaries

Statistical concepts and methods required for the time series analysis are explained. The design of an optimal controller is also discussed.

Chapter 4   A successful application

The effectiveness of the approach discussed in the preceding chapters is demonstrated with a real example of analysis and control.

Chapter 5   Computer programs

The lists of the computer programs of the statistical procedures used in the book are collected together with brief explanations. Almost all the basic procedures required for ordinary linear stationary time series analysis are included.

CONTENTS

Chapter 1

WHAT IS THE PROBLEM?

1.1 The Meaning of the Problem to Be Discussed in This Book

N. Wiener, who introduced the concept of *cybernetics* as the theory of control and communication in machine and animal, predicted that by reorganizing and unifying these theories on a statistical basis we may take a significant turn in the development toward far greater heights that ever anticipated [1]. At present, 20 years since Wiener's prediction, the validity of the prediction is being confirmed day by day. In particular, the development of the electronic computer is playing a significant role in materializing the prediction. The problem which we will here discuss may be considered to concern the realization of Wiener's prediction. The subject is how to handle statistical information to realize the control of a statistical system, using the electronic computer as a basic tool.

There is a big gap between Wiener's theory of time series and our problem. The most significant point is that Wiener's theory is based on the assumption of the availability of an infinitely long sequence of observations, while our problem stems from the fact that only finite records of observations, often of quite limited length, are available in real applications. The theory based on the concept of an infinitely long record of observations may be called a *probabilistic theory*. This type of theory must be clearly distinguished from the *statistical theory* that aims for the establishment of a connection between the probabilistic theory and observational data of finite length. The main subject to be discussed in this book then is how to develop a probabilistic representation of a real phenomenon by using statistical methods.

Another point is related to the development of electronic computers. To solve complex problems at the present time the representation of a problem in a form manipulable by computers is quite important. This point will be made clear during the discussion in the following chapters.

1

The problem to be treated in this book may thus be considered to deal with the explanation of how to realize the control of an actual system on a statistical basis, a subject which was only conceived conceptually by Wiener.

1.2   Concrete Formulation of the Problem

The data first given to us were of the type shown in Fig. 1.2-1. The figure shows the record of the gas temperature variation inside a rotary kiln used for the production of cement. As can be clearly seen an oscillation with a period of about 3 hours is dominant. The problem was how to suppress this oscillation. By looking at this figure we can quickly recognize that there are many other places where similar problems appear: an example is the control of the fluctuation of economic conditions.

To solve the problem it is necessary first to clarify why the temperature variation show a periodicity of 3 hours and what is the cause of such a fluctuation. Naturally our attention is drawn to the variation of other related variables. By observing the behavior of some concomitant variables we try to get some clue to elucidate the cause of the temperature fluctuation.

Unfortunately, in the case of the gas temperature variation, concomitant variables were not necessarily recorded on one and the same time scale. For example, the behavior of the fuel rate, which was expected to be in close relation with the temperature fluctuation, was recorded in the form of Fig. 1.2-2. Our first action was to get the records of the variables on one recording paper. An example thus obtained is shown in Fig. 1.2-3. By simply looking at the figure, we can easily recognize that it provides information that was not directly available from Fig.'s 1.2-1 and 1.2-2. Although the variables are fluctuating quite irregularly, there seems to be some definite temporal relation between their movements. Also it can be seen that oscillation of some variables are in phase and some are out of phase.

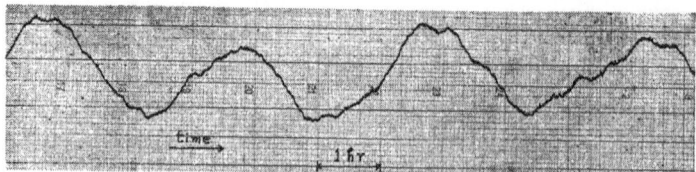

Fig. 1.2-1   An example of the record of gas temperature.

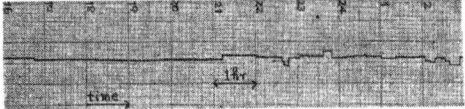

Fig. 1.2-2   An example of the record of fuel rate.

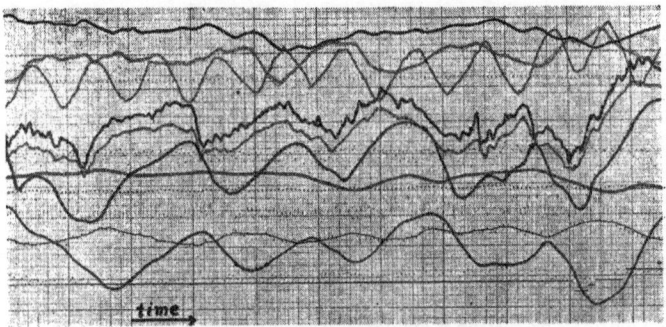

Fig. 1.2-3   An example of simultaneous recording.

The central role in any actual information processing system is played, of course, by a human observer. The importance of the data presentation in a form readily observable by the human observer can be clearly seen even by this very simple example.

1.3   Statistical Dynamic Systems

The above figure, Fig. 1.2-3, already highlights several very important problems. The first is that the behavior of the object under observation shows extremely irregular variations and, in spite of this, it maintains a certain regularity, i.e., it is a statistical phenomenon. The second is that the temporal relation among the variations of the variables has a decidedly important meaning. A system for which the temporal variation has definite meaning, and accordingly the memory or the history of the system plays an important role in describing its behavior, is called a *dynamic system*.

Examples of dynamic system can be seen everywhere. Almost every industrial production process should be considered as such a system. Also the human brain wave is known to show some definite random fluctuations. Thus, if the generating mechanism of the brain wave is considered, it should be described as a statistical dynamic system. As was mentioned earlier, the mechanism of economic fluctuations should also be properly described only as a statistical dynamic system. The behavior of the market that controls the flow of agricultural or industrial products will also be naturally understood as an example of such a system.

These observations suggest that the problem of statistical analysis and control of the cement kiln process to be treated in this book will lead, if handled in a general format not restricted by the special characteristic of the cement process, to a result that will be applicable to many other problems. Thus the discussion of the analysis and control of the cement kiln in this book will be restricted to those aspects of the handling of a statistical dynamic system that do not depend on particularly restrictive assumptions or constraints of the system.

1.4 Objectives of System Analysis

When we analyze a system which is showing complex behavior the final objective is usually the *prediction* or *control* of the behavior of the system. To attain this goal the decisive stage is the realization of a proper prediction. When a proper prediction is available, we can base our choice of action on it. When the result of our action stays within the range where it does not influence the behavior of the system, the problem remains within the realm of simple prediction, as in the case where we fasten windows based on the prediction of an impending storm. When the action produces a significant effect on the behavior of the object, this constitutes a problem of control, as in the case where restrictions on money are relaxed by the central bank to compensate for the predicted downward trend of the economy. In this case the prediction must be made on the behavior of the system operating under the influence of the control.

As was stated above, the direct objective of system analysis is the prediction and control of the behavior of the system. However, there is another important aspect, the gaining of an understanding of the structure of the system. In the case of scientific research this latter will often be the main objective. Nevertheless, in ordinary applications this aspect is often ignored and only the realization of the prediction or control is contemplated. This tendency prevents people from possible epoch-making achievements.

As will be seen from the discussion of the characteristics of statistical methods, when the data length is finite, it is theoretically impossible to determine the structure with an infinitely large number of parameters. It is necessary to adopt a model that will provide a sufficient approximation to the behavior of the system with the number of parameters kept as small as possible. For this reason it is extremely essential to constrain the range of possible structures of the model of the system by using all available information, besides that given by the record of observations.

For example, in the case of the cement kiln process, in finding a proper choice of variables for observation, it is necessary to use all the available information on the physico-chemical structure of the process and the information on the kiln behavior gained through the experience of kiln operators. This process of identification of the structure of the object is crucial for the successful realization of a control. If information processing is discussed assuming the ready availability of an accurate description of the system structure, only half the difficulty in the solution of a real problem is handled.

Generally speaking, when we analyze the behavior of a dynamic system, it is necessary to decompose the structure into two parts: the first is the source of the fluctuation of the system and the second is the structure that transforms the influence of the source into the fluctuation that confronts us. For example, in the case of the cement kiln process, the process of transporting material inside the kiln shows uncontrollable fluctuations everywhere and constitutes a source of the fluctuations of the kiln, such as the temperature and pressure variations. These fluctuations eventually build up to form the actual behavior of the kiln through a complex process of mutual interactions.

Assuming that the initial source displays some statistical characteristics we call it the *noise source*. The structure that transforms the influence of this noise source into actually observed fluctuations is called the *system structure*. To achieve the objective of statistical system analysis it is most effective to consider the fluctuation of the system as being generated by the contributions of both the noise source and the system structure. If we adopt this type of interpretation we recognize that, to achieve the final objective of control, we have two important options to take before starting the design of actual control.

The first is the handling of the noise source and the second is the modification of the system structure. It is a common mistake to expect instant success by simply implementing a ready-made control. Usually some preliminary management of the noise source and system structure is required to realized a successful design of the final control. The understanding of the physical nature of the noise source and system structure is very fundamental for this purpose. Only with the understanding of the nature of the mechanism that generates the data under observation is the utilization of the existing knowledge about the system possible. A proper understanding of the nature of the object may even lead to the suggestion of an entirely new system design.

This observation helps us to avoid the common mistakes when we approach a statistical problem. In the ordinary text book of statistical methods the choice of the optimal action is discussed under the assumption that the statistical structure of the object is known and fixed. In the text book of stochastic control the design of an optimal control is discussed by assuming knowledge of the statistical structure of the system under consideration. Except for some extraordinarily simple situations, an actual problem cannot be handled in this fashion.

It is usually the case that the statistical characteristics of the system are unknown. Also, in many cases, the structure of the system is not rigidly specified, in the sense that a detailed analysis of the structure may lead to the modification of the system structure, or to the modification of the formulation of the problem itself. This shows that it is of utmost importance to develop a proper analysis of the object and improve the environment for control, before starting any effort for the implementation of a control.

Speaking in analogous terms, the improvement of basic physiology is more important than symptom oriented theraphy to realize the ideal coordination of activities of a human body. In this sense it would be appropriate to call the action for the improvement of the system characteristics by preliminary analysis *the control of higher order*. To realize such a control all the available knowledge, or experience, or information, of the object must be fully utilized.

There is no hope of attaining the objective by a mechanistic approach. The main role of information processing of observational data is to assist the researcher in this stage of analysis. The necessity and the role of proper information processing can be grasped even by the simple example explained by Fig. 1.2-3.

When the behavior of the object shows statistical fluctuations the analysis must be done by a group of people composed of those who have specific knowledge of the object and those who are skilled in the handling of statistical data. Computers are also necessary to assist the activity of the group.

The negligence of the importance of statistical analysis forms a hindrance to success in solving a practical problem. This point is not yet generally recognized. To develop a proper appreciation of the seriousness of this observation, there is fundamentally no way but to gain some experience with real problems. However, as the second best approach, we can consider the use of real examples. It is hoped that the present book is useful for this purpose.

Here we will briefly review the characteristics of recent general theories appearing in various fields. In the application of Newtonian mechanics, when we discuss the motion of a corpuscle, the actual counterparts of those abstract concepts such as mass and acceleration can be identified uniquely. In contrast to this, in many fields there are general theories, based on statistical concepts, which do not bear much resemblance to reality. The theory only says that if it is applied to an object that satisfies the premises of the theory then it will produce a valid result. When this type of theory is applied to a new object, it is useless unless the research activity confirms the correspondence between the theory and the real object. In this book, the *statistical analysis* means the work in this stage of research activity.

1.5 How to Use Statistical Theories

The essential aspect of the problem related to a statistical object cannot be described properly without the use of some probabilistic concepts. What is first required is to develop a probabilistic model that provides an appropriate description of the behavior of the object. The information obtained during the stage of the analysis described in the preceding sections must be fully utilized for this purpose.

To specify a model the parameters within the model must be determined. Statistical methods are required to estimate the appropriate values of the parameters by using observational data. Here some statistical common sense is required to control the effect of statistical variablity of data. The same common sense is required also at the stage of the preliminary analysis.

When the actual use of a prediction or control is contemplated based on a probabilistic model, the degree of dependence of the choice of the model on the prior information must be carefully checked. Even if a procedure based on a model produced a good result in a laboratory experiment, if it is not known how to determine the structure or the parameters of the model to establish a good correspondence with the real object, the procedure is useless for practical applications.

As the amount of assumed prior information increases the uncertainty of the model is decreased and often the number of parameters to be estimated is decreased. In this situation the accuracy of the estimated parameters is increased. In the opposite situation where the number of unspecified parameters is large, a large amount of observational data is required to obtain sufficient accuracy of the estimated parameters. Thus, from the standpoint of the application of statistical methods, it is desirable to adopt a model that properly approximates the main features of the object with a simple structure that contains only a small number of unknown parameters. This fact is generally known by the name of the *principle of parsimomy* in statistics.

In the case of the fitting of an autoregressive model to be discussed later in this book a procedure is developed that can be viewed as a concrete formalization of this principle (see 3.2). The procedure also has an information theoretic justification [2]. Thus we can see that the principle of parsimony that has been developed empirically is now being given an objective formulation. This brief explanation of the principle is included here as it is a key to the successful application of statistical theories.

In this chapter some basic problems in the statistical analysis and control of a dynamic system have been discussed. The lengthy verbal general discussion may not have been very helpful in developing proper understanding of the problems. Hereafter we will concentrate on real examples. A minimum amount of necessary statistical concepts is explained in Chapter 3.

REFERENCES

The references [3] by K. J. Aström and [4] by D. M. Himmelblau are closely related to the subject treated in this book. The proceedings [5] of a conference on multivariable control systems might also be of interest to readers of this book. References that are related to the general problem of model construction are [6], [7] and [8].

[1] N. Wiener, *Extrapolation, Interpolation and Smoothing of Stationary Time Series*, John Wiley & Sons, New York (1949).
[2] H. Akaike, Information theory and an extension of the maximum likelihood principle, In *2nd International Symposium on Information Theory*, B. N. Petrov and F. Csaki, eds., Akademiai Kiado, Budapest (1973) 267–281.
[3] K. J. Åström, *Introduction to Stochastic Control Theory*, Academic Press, New York (1970).
[4] D. M. Himmelblau, *Process Analysis by Statistical Method*, John Wiley & Sons, New York (1970).
[5] H. Schwartz ed., *Multivariable Control Systems*, Vols. 1–4, North Holland, Amsterdam (1971).
[6] M. G. Kendall, Model building and its problems, *Mathematical Model Building in Economics and Industry, First Series*, Charles Griffin, London (1968) 1–14.
[7] R. J. Ball, Econometric model building, *Mathematical Model Building in Economics and Industry, First Series*, Charles Griffin, London (1968) 15–30.
[8] A. W. Phillips, Models for the control of economic fluctuations, *Mathematical Model Building in Economics and Industry, First Series*, Charles Griffin, London (1968) 159–165.

Chapter 2

AN EXPLANATION OF THE CONTROLLER DESIGN PROBLEMS

In this chapter the difficulty of controlling a statistical dynamic system is explained by using the real example of a cement kiln process. The purpose of this chapter is to clarify the need for the type of method to be discussed in Chapters 3 and 4. Readers may skip this chapter without adversely affecting the reading of the following chapters.

2.1 What is a Rotary Kiln?

A cement rotary kiln is a tilted rotating cylinder with a diameter of 4–5 m and length of 40–200 m, as shown in Fig.'s 2.1-1 and 2.1-2. The original raw material is fed through the higher end of the kiln and undergoes heat exchange during the passage through the zones for drying, preheating, calcining and burning. The material is then discharged from the lower end of the kiln into the clinker *cooler*. The raw material, which is a mixture of limestone, silica, clay, and iron sludge, stays inside the kiln for a certain period of time; for example, 40–50 minutes or above 3 hrs, depending on the degree of tilting, length, and rotation speed of the kiln and the condition of the wall coating. During this period the raw material completes the endothermic and exothermic reactions and undergoes the burning process to form clinkers.

Here the chemical reaction taking place inside the kiln of the "wet process" type is explained briefly. There is also another type of process, called the "dry process", where the material does not pass the state of slurry and the process starts at stage 2 of the following description of the wet process. The differences of the two processes lie mainly in the structure of the drying zone and there is not much difference in the overall characteristics.

9

Fig. 2.1-1   External view of rotary kilns.

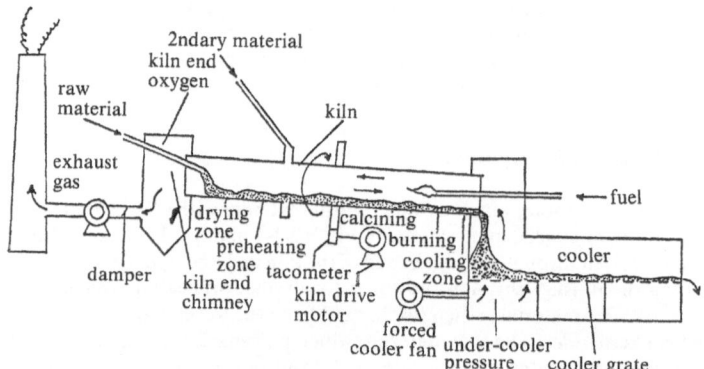

Fig. 2.1-2   Schematic diagram of a rotary kiln.

1) slurry (20°C) $\xrightarrow{H_2O}$ dry material (100°C)   : drying zone
2) dry material (100°C) $\longrightarrow$ preheated material   : preheating
                                         (about 900°C)   zone
3) $CaCO_3 \xrightarrow{900-1000°C}$ $CaO+CO_2$: calcining zone
    $MgCO_3 \longrightarrow$ $MgO+CO_2$ (endothermic)

4) $2CaO + SiO_2 \xrightarrow{1480°C\sim}$ $(CaO)_2 \cdot SiO_2$: burning zone
     $CaO + (CaO)_2 \cdot SiO_2 \xrightarrow{\,''\,}$ $(CaO)_3 \cdot SiO_2$ (exothermic)
     $3CaO + Al_2O_3 \xrightarrow{\,''\,}$ $(CaO)_3 \cdot Al_2O_3$
     $4CaO + Al_2O_3 + Fe_2O_3 \xrightarrow{\,''\,}$ $(CaO)_4 \cdot Al_2O_3 \cdot Fe_2O_3$

5)   clinker $(1500°C) \xrightarrow{cooling}$ clinker $(1200°C)$: cooling zone

6)   clinker $(1200°C) \xrightarrow{cooling}$ clinker (room temperature)   : cooler

A process with this type of kiln is technically known as a counter flow type distributed delay heat exchanger. The control of this seemingly simple kiln process exhibits typical difficulties usually encountered in handling a complex system.

2.2 Control of the Kiln

The variables of interest and their measuring points in a kiln are illustrated in Fig. 2.2-1.

In the figure, T, T_{ig}, T_{bz} and T_2 respectively denote the kiln end gas temperature, intermediate gas temperature, burning zone temperature, and secondary air temperature, which is the temperature of the air being blown into the kiln from the cooler. These four kinds of temperature are typical variables that represent the heat conditions inside the kiln. P_e denotes the kiln end draft pressure, P_h the draft pressure inside the hood at the lower end of the kiln, P the under-cooler grate pressure. These three kinds of pressure represent the draft pressure distribution inside the kiln. O_2 denotes the content

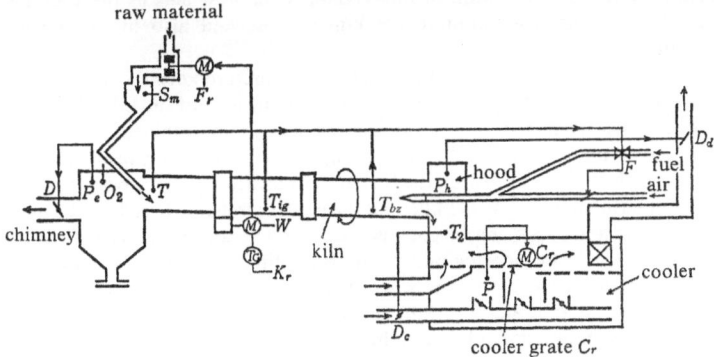

Fig. 2.2-1   Process variables and their measuring points.

of oxygen in the kiln end exhaust, W the load power of the kiln drive motor, K_r the kiln rotation speed, F the fuel rate, C_r the cooler grate speed, D_c the damper opening of the cooler, D the kiln end exhaust damper opening, D_d the opening of the damper that releases the excess air blown into the kiln from the cooler, F_r the flow meter reading of the raw material, and S_m the weight measurement of the raw material being fed into the kiln.

2.2.1   The first form of control: the human operator

As was mentioned earlier, the kiln process treats the material in the form of slurry or powder. The raw material fed into the kiln is first dried and then, after being heated to 900°C or above, fed into the calcining zone where calcium carbonate, which is one of the main constituents of the raw material, is decomposed into quicklime and carbon dioxide by absorbing heat. At this point the air slide phenomenon occurs, which is the increase of the flow rate of the powder caused by the gas discharge from the material. However, owing to the temperature fluctuations, the conditions for the carbon dioxide generation vary and this causes the variation of the material flow rate.

The material is further heated and starts emission of heat in the burning zone. Here the sticky material forms an irregular coating on the inner wall surface of the kiln. The coating is formed randomly and often grows into a particularly obtrusive form, called a mud ring. This mud ring disturbs the material flow and sometimes a rush of material is caused by its sudden collapse. The material flow at the lower end of the kiln is composed of a succession of clinker with varying thermal history and is fed into the clinker cooler to be quickly cooled down by the air flow during the passage over the cooler grate which transports clinker by its reciprocal movement. The air that absorbed heat from the clinker flow is called the secondary air and is blown into the kiln. The variation of the secondary air flow acts as the source of variation of the heat input to the kiln process and also of the burning conditions.

In the operation of the kiln process it is desired to keep the material and gas temperature distribution stationary along the axis of the kiln so that the thermal history of the material is homogeneously maintained. The human operator watches a part of the burning zone through the peek hole and controls the kiln by adjusting the fuel and/or air flow rate. However, dust is densely suspended inside the kiln and the control must be based on the quick perception and experience of the operator. As a result of this type of control precipitation of the coating and the firebrick lining of the kiln occurs quite often. At one time there was a tendency to assume such troubles to be inevitable and consider the kiln an uncontrollable monster. This was a common phenomenon observed in many areas of the chemical process industry in those days.

In any industry based on some kiln process, such as the steel or glass producing industry, the kiln was usually given a nickname depicting it as a monster. This phenomenon is not limited to the chemical process industry. It is a typical reaction of people who are faced with a system which is complex and difficult to handle.

Even under such difficult conditions, operators were making efforts to realize homogeneous burning of material by using the observation of some physical characteristics of the clinker, such as shape, gloss, brightness, and hardness. It was obvious that if only stationary operation of a kiln was realized this would directly lead to the production of clinkers with homogeneous quality. Kilns in those early days were smaller compared with those of the present and response to manipulation was speedy. Since these kilns usually tended to return to stationary operation, operators tried only to bring kilns into stationary operating condition and then left them untouched until significant disturbances took place. However, because of the increasing demand for cement production and the movement for the reduction of the cost of labor the kiln size was increased to the point where quick response to manipulation could not be expected, thus making operator's empirical corrective action not quite sufficient to stabilize the behavior of the kiln.

To counter this difficulty, instrumentation was improved to collect information about the internal behavior of the kiln and local automatic control loops were formed to try to reduce fluctuations of such variables as temperature, pressure etc., at each part of the kiln, without paying attention to the behavior of variables at other parts.

2.2.2   The introduction of local regulator control

Local regulator control is a control that tries to keep a variable such as temperature, pressure and flow rate at a fixed value, called the set point. Any production process is exposed to various sources of disturbance; for example, the variation of the calorific content of fuel, variation of the environmental temperature of the process, and so forth. A local regulator control aims at keeping one of these process variables at the set point despite those disturbances. The control is realized by measuring the deviation of a variable from its set point and then applying an operation to cancel the deviation. This type of control that returns the error of control back to the control operation is called a *feedback control*.

The main technique developed for the local regulator control was the so-called PID-control. P, I, and D are respectively the abbreviations of proportional, integral and differential and represent the characteristics of a particular control. A control system must be stable, but also it must smoothly cancel the error, the deviation of the controlled variable from its set point, as quickly as possible. The control that feeds back the operation that is proportional to the error is called P-control and the one that feeds back the operation that is proportional to the integral of the error is called I-control. When the amount of feed back is proportional to the differential of the error it is called D-control.

A combination of the above three types of control is called a PID-control. I-control is effective in compensating for slow variations of the control error and D-control for rapid variations. However, commercially available PID-controllers were usually designed for general use and their parameters had to be adjusted for each particular application. In the case of the cement kiln process, the most effective and simply realizable control is that of the feed rate or water content of the raw material. The implementation of the control of a kiln started with the control of this type of variable that is independent of the internal process of the kiln.

Here and in the following Section 2.2.3, the effectiveness and the limitations of the local regulator control of the kiln process will be discussed. The operating conditions of a kiln process can typically be classified into the following three types:

1) Stationary operating condition (see Fig. 2.2.2-1)

This is the situation where the axial distributions of the gas and material temperature respond proportionally to the control input. In this case the response to the control input is clearly visible and the effect can easily be confirmed. For example, when the fuel rate is increased the temperature rises at both ends of the kiln, and vice versa. Thus the variable responds to control similarly at both ends and stationary material flow is maintained.

2) Stationary repetition of locally non-stationary behavior (see Fig. 2.2.2-2)

This state often appears when the amount of production is increased up to a certain point under human operator control. Here the mutual interference

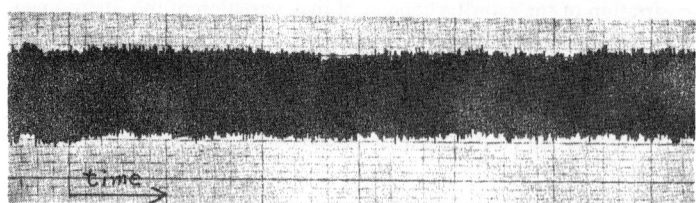

Fig. 2.2.2-1 An example of the kiln drive motor power variation recorded in a stationary condition.

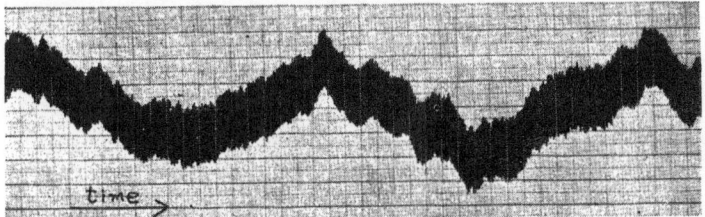

Fig. 2.2.2-2   An example of the record of kiln drive motor power variation showing
stationary oscillations.

between variations of the state at various points of the process becomes
non-negligible.

For example, consider the situation where the lower end temperature has
dropped abnormally due to some disturbance of the process and the increase
of fuel rate is being contemplated. In this case as the lower end temperature
recovers the kiln end gas temperature rises. Thus when the raw material is
transported from the upper end to the lower, its temperature is higher than
normal. However, since the raw material at the lower end is in its final stage of
burning, this rise of temperature must definitely be avoided. Accordingly the
reduction of fuel rate becomes mandatory. As a consequence the temperature
shows wavy fluctuations in the axial direction of the kiln, with the
temperature kept stationary at the lower end. The transport of material is
affected by the fluctuation of the temperature distribution and the overall
behavior of the kiln shows a pattern of repetition of stationary oscillations.
3)   Unstable significantly non-stationary behavior (see, Fig. 2.2.2-3)

This is caused by a sudden change in the quality or mass of the raw
material. It is also often caused by a disturbance of the internal transport of
material, often observed when the kiln is operated aboved its rating.

When an unevenness appears within the transport process the locally
dense portion of the material shows increased fluidity due to the low heat
exchange rate. This causes sliding which pushes the raw material down
through the lower end of the kiln. When the excessive raw material flows into
the cooler grate it causes an increase of the resistance against the forced air
flow through the cooler. This excessively reduces the amount of air flow into
the kiln and lowers the combustion rate of the fuel. Thus the temperature of
the raw material at the lower end is further reduced and this causes a positive
feedback that causes divergent behavior of the kiln.

As is explained above the actual behavior of a kiln process can be

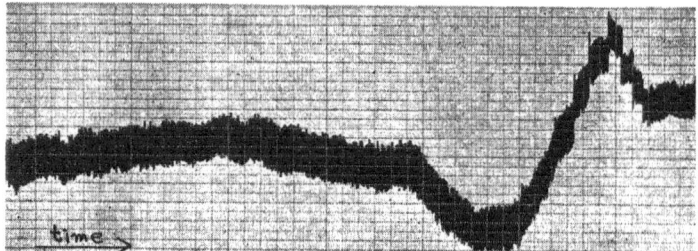

Fig. 2.2.2-3   An example of the record of kiln drive motor power variation showing divergent behavior.

classified into several states and as can be seen from cases 2) and 3) the variation of the process is first caused by some local disturbance. Accordingly, local regulator controls were introduced at those places where disturbances tend to occur and at the same time the implementation of some control was possible.

2.2.3   Local regulator control in practice

Before the introduction of computer control, the following local controller configuration was commonly adopted for the kiln process (see, Fig. 2.2-1).

1)   Stabilization of under-cooler grate pressure by adjusting the cooler grate speed.

2)   Hood draft (the pressure at the joint of the cooler and kiln) control through the adjustment of the cooler exhaust damper opening.

3)   Stabilization of the secondary air temperature through the adjustment of the cooling fan damper.

4)   Stabilization of the kiln end draft (or content oxygen) by adjusting the kiln end damper.

5)   Stabilization of the kiln end gas temperature, intermediate gas temperature, and burning zone temperature through the adjustment of the fuel rate.

Each of these controllers is a local feedback controller for a single input and single output and the kiln control has usually been realized by a multi-loop system composed of some combination of these controllers. The regulator control of a local system can easily be realized by a single input and single output control, if the system is isolated. However, when there are other control systems operating simultaneously, there arises the possibility of instability due to the interaction between control systems. Accordingly, in the case of the kiln process, it was customary to first apply a regulator control to a subsystem that was independent of, or only weakly coupled with, other systems.

Here we will discuss problems in the practical application of parallel combinations of controllers which were tried in the cement kiln process. The mathematical explanation of the interference phenomenon common to these examples will be given in 2.3.2.

Fig. 2.2.3-1 gives the diagram of a real example. This control was designed to adjust the fuel rate (F) so as to keep the kiln end gas temperature (T) at a fixed level and simultaneously to adjust the kiln end gas damper opening to realize the draft control. The control action simulates the basic manual control of a human operator. When there is a disturbance in the water content or mass of the raw feed material this naturally causes variation in T. For example, when water content is increased T is reduced. Then, if the fuel rate is increased to bring T back to its normal value the oxygen content O_2 of the air inside the kiln decreases. If the damper opening is increased to bring O_2 back to its normal value the kiln end gas temperature T increases, due to the increased air flow.

At first sight, these two control loops appear to be supporting each other. However, in actual operation, the behavior of one loop causes overreaction of another and, because of the temporal relation between the two loops, the process often starts oscillatory variations. In particular, the effect of fuel control produces a definite effect on the oxygen content and reveals the importance of the problem of interference between parallel control loops.

As the next example, let us consider the cooler control. Fig. 2.2.3-2 depicts the cooler control system. This system aims at controlling the under-cooler grate pressure (P) and secondary air temperature (T_2) by manipulating

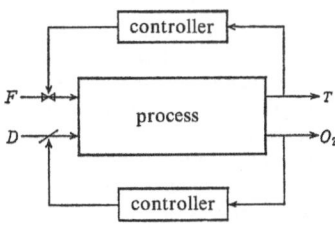

Fig. 2.2.3-1   An example of parallel regulator control.

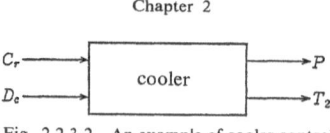

Fig. 2.2.3-2   An example of cooler control.

the cooler grate travel speed (C_r) and forced fan damper opening (D_c), respectively. These two control loops also show significant interaction. For example, if the cooler grate speed is adjusted to keep the under-cooler grate pressure fixed this causes a variation in the secondary air temperature that flows through the cooler grate into the kiln. When the forced fan damper opening is adjusted to counteract the temperature variation it causes variation in the under-cooler grate pressure. Thus the whole control system frequently becomes oscillatory like a pendulum. As this explanation shows, even for the separate cooler system, the realization of control of a multiple input-multiple output system is quite difficult. One can easily guess that the successful control of a multiple input-multiple output system is beyond the scope of the simple application of a PID controller.

The above examples show that a collection of local regulators require a higher order control that coordinates the subsystems. This suggests the necessity of introducing a computer controller that can easily realize the required complicated response characteristics and decision mechanisms. Before finally going to a discussion of computer control let us further briefly review the characteristics of the human operator control.

2.2.4   Characteristics of the human operator within a control system

With a process that shows very low frequency fluctuations, such as the gradual drift of operating conditions often observed in large capacity chemical process systems, the human operator is usually too much occupied by other problems at hand to pay sufficient attention to the long range behavior of the system. Also, because of the operator's change, the memory of past observations of the process is discarded. Thus the adaptability and flexibility that are the distinguished characteristics of the human operator usually make him rather near-sighted and forget the importance of the *dc* (*direct current*) *component* that is basic to the process. The regulator control discussed in the preceding section was required mainly to compensate for this shortcoming of human control.

Further, actions taken by a human operator under similar circumstances are not always identical. This means that operators are unkonwingly generating disturbances by unnecessary operations. Actually, in the case of the cement kiln process, the good reputation of an operator is sometimes due to the way he handles disturbances generated by human operation.

Nevertheless, once a difficult situation arises, human operator can handle it through skillful use of process signals based on his long experience. In this sense, the regulator control that mechanically tries to keep each variable at a fixed level is definitely inferior to the human control. In spite of this, it is quite desirable to suppress the noise generated by the human operator and here again we see the necessity of the transition to computer control.

2.3 Early Computer Control

As was mentioned in the preceding section, the requirement of the production process which was not fulfilled by conventional control based on PID controllers demanded the introduction of the computer into process control. In the cement industry, the first computer control system was introduced for the dry process control of the Orogrande factory of the Riverside Cement Company, U.S.A., in 1959. This system used the RW 300 control computer developed by the Ramo-Wooldrige Company and was designed for the control of the blending and kiln processes [1, 2]. The same computer was introduced to the Kumagaya factory of the Chichibu Cement Company, Japan, in 1962. This was the first application to the wet process [3]. Fig. 2.3-1 gives the picture of the system. In the following sections, we will historically review the development of computer control at that time.

Fig. 2.3-1   View of the instrument panel and the computer console (RW 300) of one of the earliest computer control systems of the cement process.

2.3.1   Mathematical model

When a computer is used for the purpose of process control, the development of some mathematical model forms the starting point. In the case of the rotary cement kiln process there is an approach that treats the process as a distributed parameter system and develops the model solely by mathematical analysis of the system [4, 5]. However, the construction of the theoretical model is based on physico-chemical analysis and is not free from the serious limitation of presupposing numerous assumptions. This is a phenomenon commonly observed when a mathematical model is developed for a complex system by some theoretical analysis and is not limited to the case of the kiln process.

By the physico-chemical approach to the model the construction of the model is centered around the heat and mass balance. An example is given by a system of 18 simultaneous first order partial differential equations and 4 algebraic equations, with corresponding initial and boundary conditions. These equations depend on the physical shape of the kiln and the numerical analysis of the system is quite elaborate [5]. In this case, although the model is called theoretical, it contains numerous unmeasurable parameters, such as the heat conduction rate of the raw material, that of firebricks of the inside wall, the specific heats of these materials and the gas, etc. Values of these parameters are chosen so that the numerical analysis produces a reasonable result. The model thus obtained is usually appropriate for the static and qualitative description of the kiln and is useful either for the modification of the basic design of the kiln or for the determination of set points or dc reference values of process variables. However, because of the static nature of the representation and the limited accuracy, the model is almost inapplicable to the control of the actual kiln operation. For the purpose of control, what is really required is a model that provides a good representation of the dynamic behavior of the process. Here we will briefly consider the construction of the behavioral model for this purpose.

There is an approach to modeling where the modeling is mainly based on the relation between the variations of the input and output of the system, rather than on other physical characteristics such as the mechanical structure and shape. In this approach, attention is concentrated only on how the input to the system is transformed into the output, i.e., it concentrates on the aspect of the transmission of signals.

For the description of the relation between a pair of input and outputs the concept of transfer function is useful. The *transfer function* of a *linear time invariant system* is defined by the Laplace transform of the impulse response function of the system. When the Laplace transform of the input $x(t)$ is given by $X(s)$ and the transfer function of the system is given by $G(s)$, the Laplace transform $Y(s)$ of the output $y(t)$, is given by the relation $Y(s)=G(s)X(s)$, where the *Laplace transform* $G(s)$ of a function $g(t)$, with $g(t)=0$ for $t<0$, is by definition

$$G(s) = \int_0^\infty e^{-st} g(t) \, dt .$$

A real process is composed of many variables. Accordingly a behavioral model based on transfer functions is given by a set of these functions. The parameters that determine the transfer functions are chosen to approximate the response of the real process. To test the behavior of the model the input must be specified. Commonly used test inputs are the step and sinusoidal function and the corresponding response is called the *step* and *frequency response*, respectively. However, there are some problems for the application of these test inputs to an actual process. These are

1)   the possible change of the system characteristics due to the disturbance caused by an excessively large input, and

2)   the low signal to noise ratio when the input is kept small, and

3)   the general difficulty of applying an artificial input to an actual process operating in a production line.

The application of a test signal is thus usually limited to the situation where the process is operating without generating much noise. Thus the test is often very difficult for a process with a very large time constant, such as the kiln process, where faster variations of other variables are not completely controllable and generate significant noise through a complex process of mutual interactions.

Consider the response of the clinker temperature to the step input of fuel rate. In this case the response of the clinker temperature is composed of two components, the direct response by heat conduction to the change of fuel rate and the response through the change of the thermal conditions of the whole kiln system. The *settling time* of the response is quite long, but the internal noise, such as the uneven flow of material, and the fluctuation of other variables makes the indefinite continuation of the test impossible.

The difficulties may be considered as manifestations of the incompleteness of the approximation by the linear time invariant model. Nevertheless, let us temporarily ignore the difficulty of modeling and consider the design of a control system based on a model defined by a collection of single input-single output systems.

2.3.2   Controller design

To apply the design concept of the controller developed for a single input-single output system to a multivariable system it is necessary to introduce a *compensation circuit* to eliminate the interference between variables discussed in 2.2.3. By this approach the multivariable system is expected to be transformed into a collection of mutually independent single input-single output systems. The design procedure for the controller of a single input-single output system that could be developed by depending on the designer's intuition cannot be extended to a multivariable system without proper handling of the problem of interference.

Here we will consider a relatively simple example of the two input-two output process previously illustrated in Fig. 2.2.3-2. By using transfer functions this system is represented by Fig. 2.3.2-1. In this figure, $G_p(s)$ and $G(s)$ represent the transfer function matrices of the cooler process and controller system, respectively. By the transfer function method, the Laplace transform of the system output is obtained by multiplying the Laplace transform of the system input by the transfer function of the system. The effect of a linear time invariant system is thus represented by a simple multiplication operation.

In the two-variable system of Fig. 2.3.2-1 the compensating elements are represented by G_{C1} and G_{C2}, and the regulating elements by G_{R1} and G_{R2}, where G_{R1} transforms X_1 into the forced cooler fan damper position D_c and G_{R2} transforms X_2 into the cooler grate speed C_r. Denote the deviation of the secondary air temperature Y_1 from its set point by X_1 and the corresponding deviation of the under-cooler grate pressure Y_2 by X_2. Then it holds that

$$Y_1 = X_1(G_{R1}\, G_{11} + G_{C1}\, G_{21}) + X_2(G_{C2}\, G_{11} + G_{R2}\, G_{21})$$

$$Y_2 = X_2(G_{R2}\, G_{22} + G_{C2}\, G_{12}) + X_1(G_{C1}\, G_{22} + G_{R1}\, G_{12})$$

(1)

Here the transfer functions G_{ij} ($i, j = 1, 2$) represent the characteristics of the process. The argument s of the Laplace transform is suppressed when it does

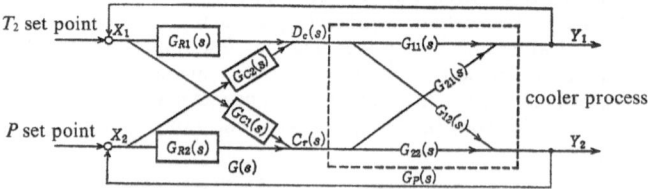

Fig. 2.3.2-1   Diagram of a cooler controller (2-input 2-output system).

not cause any confusion. In equation (1), Y_1 and Y_2 are controlled by X_1 and X_2. The second terms on the right hand sides of (1) represent the interaction. In these terms, if the coefficients of X_2 and X_1 are put equal to zero the two control loops become free from interaction.

For this purpose we have only to put

$$G_{C1} = -\frac{G_{R1} G_{12}}{G_{22}}$$

$$G_{C2} = -\frac{G_{R2} G_{21}}{G_{11}}$$

(2)

In this case, since we have $Y_1 = T_2$ and $Y_2 = P$, we get

$$T_2(s) = X_1 \cdot G_{R1}\left(G_{11} - \frac{G_{21} G_{12}}{G_{22}}\right)$$

$$P(s) = X_2 \cdot G_{R2}\left(G_{22} - \frac{G_{12} G_{21}}{G_{11}}\right)$$

(3)

and the under-cooler pressure P responds only to $X_2 G_{R2}$ (=cooler grate speed C_r) and the secondary air temperature T_2 to $X_1 G_{R1}$ (=forced cooler fan damper position D_c). If $G_{11}(s)$, $G_{12}(s)$, $G_{21}(s)$ and $G_{22}(s)$ can be obtained by test operation of the process the characteristics of the necessary compensation circuit can easily be obtained by simple calculation. One of the big merits expected of computer control was the simplicity of realizing approximations to elements required by this type of design.

2.3.3 Construction of a control system

At the earlier stage of the introduction of computer control, the process was approximated by a set of transfer functions obtained by testing, as was described in 2.3.2. The basic approach to control system design was first to develop a feedback control for each process variable by using the relation between the variable and an input and then to add the compensation element to eliminate the effect from other variables. Since the transfer function representation of the process characteristics can easily be understood, the design of the control system was developed mainly by using diagrams and simple calculations with the help of an intuitive understanding of the process.

This was the royal road to the controller design when the design of a servo-system for the control of a mechanical system with a faster response than the industrial process was of main interest. Accordingly, at the initial stage of the introduction of computer control, the users were mainly concerned with computerization of conventional process analysis and controller design, digitalization of analog PID controllers, compensation for the process delay, and realization of arbitrary response characteristics, i.e., they concentrated on exploiting the merits of a digital computer in the form of so-called DDC (direct digital control). Fig. 2.3.3-1 shows the basic structure of computer control at that time. The actual content of the figure is illustrated in Fig. 2.3.3-3.

The construction of the controller of Fig. 2.3.3-1 is given in Fig. 2.3.3-2. Here $D_i(z)$ and $G_C(s)$ represent elements that are realized by the computer and its peripheral equipment.

With the introduction of the digital technique, intermittent discontinuous control was advanced, where the Laplace transform was replace by the z-transform. The *z-transform* of the function $x(t)$ measured at $t=k\Delta t$ $k=0, 1, 2, ...$) is defined by

$$X(z) = \sum_{k=0}^{\infty} x(k\Delta t)\, z^{-k}.$$

For the use of the z-transform in control system analysis and design readers are referred to the book by Jury [7].

Fig. 2.3.3-3 gives the final structure of the computer control system of the cement kiln process. How this control system worked in practice and what remained as problems will be discussed in the next section.

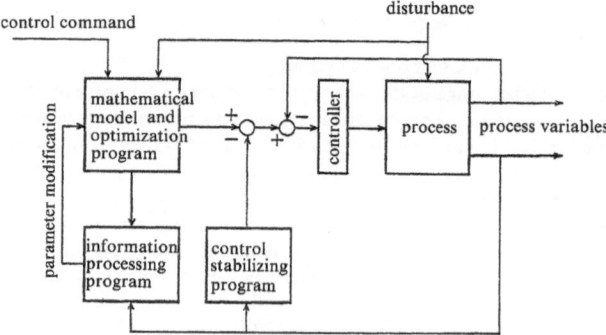

Fig. 2.3.3-1   The basic structure of the computer control.

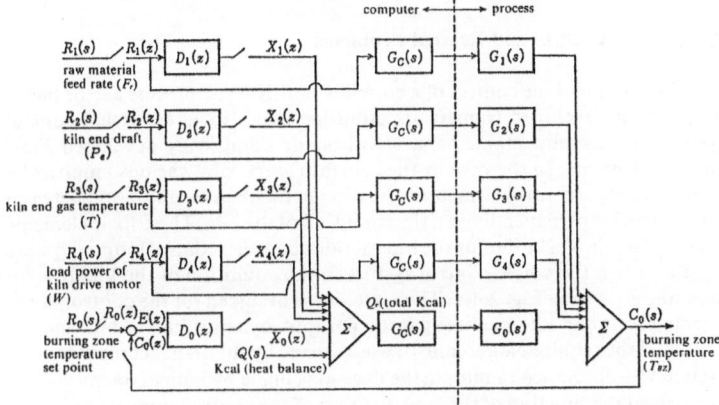

Fig. 2.3.3-2 The diagram of the computer control of fuel rate.
$R_i(s)$: set points given by the model. $G_i(s)$: process transfer functions. $D_i(z)$: transfer functions of the compensating elements. $G_C(s)$: sample hold circuits. $G_0(s)$: main process output variable.

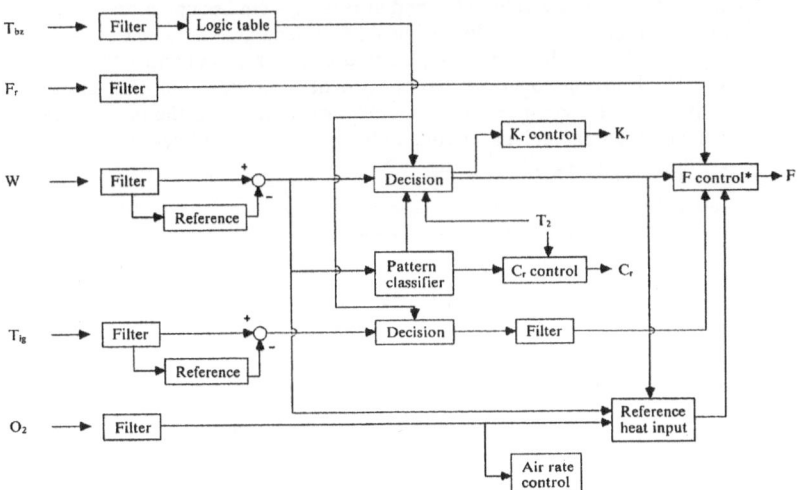

*F control is realized by a combination of burning process model, optimization and model adjustment programs and a program for stabilization.

Fig. 2.3.3-3 An example of a computer control diagram. Each filter is realized by a linear time invariant system. For the explanation of the abbreviated notations see 2.2.

2.4 Actual Control and Related Problems

The design of the control of a complex multivariate process can be based
on the well developed transfer function method. However, actual control
systems are usually supplemented with some empirically developed logic
control elements. In the case of the cement rotary kiln, various empirically
developed logic control elements were used to realize a coordination between
the control loops, depending on the condition of the kiln. These logic elements
played the role of a buffer for the interactions between control loops and were
used to adapt the whole control system to the condition of the kiln [6]. An
example is given in Fig. 2.4-1. The system was designed for the control of the
burning zone and intermediate gas temperature by adjusting the kiln end gas
draft and the required logic control was realized by the digital computer. This
system was developed to imitate the decisions made by human operators and
performed the function of the coordination of the control of the two kinds of
temperature.

At the early stage of the introduction of computer control, it was
observed that a control system was more susceptible to random disturbances
when it was designed by an elaborate application of conventional techniques
based on the mathematical model, and also the system became more robust
when the number of empirically designed logic elements was increased. This
circumstance lead to the somewhat passive use of non-linear control that was
activated only when a big disturbance brought the system into an unstable
condition. Control engineers were mainly concerned with the design and
adjustment of empirical logic circuits to be used in the non-linear region.

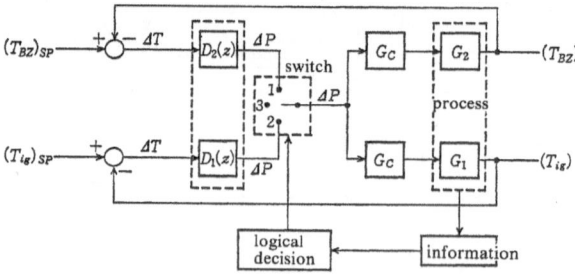

Fig. 2.4-1   An example of empirically designed logic control.

Some examples of kiln operation records under these circumstances are given in Fig.'s 2.4-2, -3, and -4. Fig. 2.4-2 demonstrates stationary operating conditions. Fig. 2.4-3 compares the action taken by the kiln operator with the dry run output of the control computer. The kiln was in the state of transition from the non-stationary to the stationary condition. The adjustment of the computer control was performed through the observation of this type of result. Fig. 2.4-4 is an example that shows the difficulty in the actual application of computer control. It shows that the disturbance of W, the kilowatt kiln drive motor load, that occurred at some time point continued to increase the amplitude until the computer control was cut off.

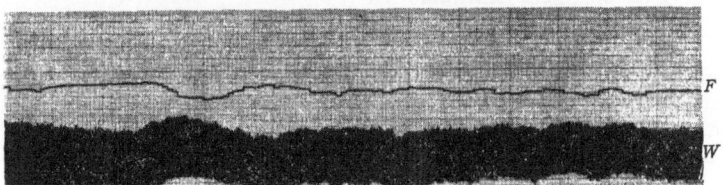

Fig. 2.4-2   An example of the record of a kiln computer control (stationary operation).

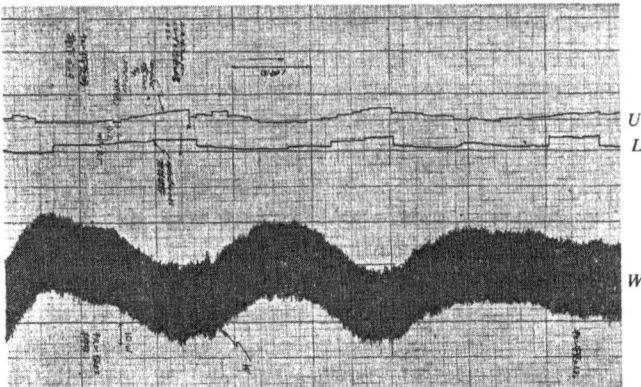

Fig. 2.4-3   Comparison of fuel control by an operator (L) and the direction from the control computer (U).

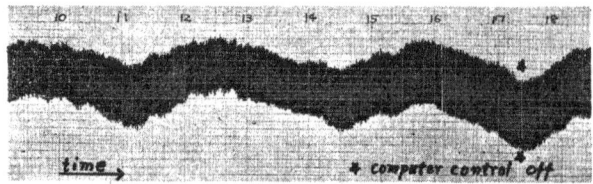

Fig. 2.4-4   An example of discontinued computer control.

2.5   Overcoming the Difficulties

To overcome the difficulties in controlling the cement kiln discussed thus far, we have to pay our attention to several points soon to be discussed. A computer control realization based on the consideration of these points will be discussed in Section 4. Related statistical methods are discussed in Section 3. To develop a general understanding of the problem, the reader may proceed directly to Section 4 after finishing this section.

The points to be considered are:

1)   To understand the kiln as a statistical dynamic system

As can be seen from the preceding descriptions, a clear recognition of the fact that the system shows a statistical fluctuation was completely missing in the conventional controller design. This is a fatal fault. For the evaluation of the efficiency of a control system the condition of the environment under which the controller operates must first be taken into account. The lack of recognition of the fact that the system shows statistical fluctuations means a basic criterion for the evaluation of desired operation of the system is also lacking. The understanding of the kiln as a statistical dynamic system constitutes the starting point for overcoming these difficulties. The recognition of this point leads to an understanding of the importance of the following three points.

2)   To confirm the statistical characteristics of the system

The structure of the kiln and the characteristics of the noise in an actual operating condition can be confirmed, up to certain point, by the statistical method to be described in the next section. It is necessary to develop a unifying objective understanding of the behavior of the system by properly coordinating the information thus obtained with empirical observations and conventional theories. Only by this approach can we obtain an objective view of the causal relations that governs the behavior of the kiln.

3)   Handling of a complex system by the state space representation and dynamic programming

The difficulty of controller design for a multivariate system can be reduced drastically if the control theory of a dynamic system based on the state space representation is adopted and the optimal controller design is performed by computer using the dynamic programming approach.

4)   Improvement of the instrumentation

In the preceding discussion, the difficulty of measuring the necessary variables was not mentioned, for simplicity. For example, the difficulty in measuring the burning zone temperature, or the oxygen content of the kiln end gas, or the secondary air temperature is so significant that at present no measurement of these quantities is available in a form that is directly useful for the purpose of control. This constitutes a serious obstacle in confirming the system characteristics or in the realization of control. To overcome this particular difficulty, it is necessary to utilize empirically obtained information as much as possible for the elimination of the noise in the measurement, and also to improve the basic standard instrumentation. Statistical processing must be performed when necessary.

Now, at last, we are at the starting point of our expedition to overcome the difficulties, and to understand the behavior of the kiln, as described in 1) above.

REFERENCES

[1]   J. M. Sauer, L. W. Weeks, P. T. Fleeman and V. A. Kaiser, Riverside puts kiln under closed-loop computer, *Pit and Quarry*, July, 1962.

[2]   J. W. Lane, Ousting human error, computers boost kiln output, decrease costs, *Rock Products*, Sept., 1962.

[3]   E. Tashiro, T. Nakagawa, J. Hasegawa, Chichibu Cement Co's new plant at Kumagaya is now under real time computer control, *Proc. International Symposium on Application of Statistics, Operations Research and Computers in the Mineral Industry*, Denver, U.S.A., April, 1964.

[4]   R. Stillman, A paper presented at A. I. Ch. E. National Meeting, Calif., May, 1965.

[5]   R. Stillman, Cement kiln simulation using oxide chemistry, *IBM Technical Report*, June, 1964.

[6]   T. Nakagawa, An example of the numerical control of an inverse response process, *Keisoku to Seigyo* (*Instrument and Control*), vol. 2 (1963) 952–956 (in Japanese).

[7]   E. I. Jury, *Sampled-Data Control Systems*, John Wiley, New York, 1958.

Chapter 3

STATISTICAL PRELIMINARIES

In this chapter, explanations will be given of some statistical concepts and methods that are necessary for the understanding of the discussion in the following chapters. First, in Section 3.0, some basic concepts of probability theory which play fundamental roles in describing statistical phenomena will be discussed. Those who are already familiar with probability theory and statistics in general may skip this section. Also, since the use of the concepts and methods to be introduced in this chapter will be explained with examples in Chapter 4, readers who are less interested in technical details may proceed directly to that chapter. Technical details are further summarized as computer programs in Chapter 5. For a proper use of these programs the understanding of the content of this chapter is necessary. However, by experimenting with these programs using real examples, the reader can easily develop a proper understanding of the content of this chapter. Within the text, Program 5.3.2 etc. indicates the corresponding computer program of Chapter 5. As a general reference for this chapter, the reader is referred to [1].

3.0 Probabilistic Concepts

An appropriate description of a phenomenon generated by a so-called random mechanism can only be obtained with the aid of probabilistic concepts. In fact, it is impossible to develop a reasonable discussion of the problems of statistical analysis and control, which form the subject of this book, without the use of some stochastic concepts. In this section we will discuss the minimum amount of concepts required for the discussion in the following chapters.

3.0.1 Probability

For each experiment or observation, we can think of the probability of occurrence of a certain event. To define probability explicitly it is necessary to give an appropriate mathematical expression for this event. For this purpose, we consider a space (*label space*) consisting of a set of labels which are necessary and sufficient to uniquely represent the outcome of each experiment. Probability can formally be defined by distributing one unit mass over this label space.

A *label* which is an element of the label space is called an *elementary event* and is used to represent a basic outcome of the experiment. Generally, an *event* can be represented by a set of all the elementary events or outcomes that have the properties that characterize the event. Probability is defined for this type of set of elementary events. An elementary event is represented by ω, and the label space, the set of all ω's, is represented by Ω. An event is then described by an appropriate subset A of Ω. If we let F stand for the set of all possible events, then the probability $p(A)$ can be defined as the value given to A by the function $p(\cdot)$ that assigns a number between 0 and 1 to each element of F.

One may interpret the probability $p(A)$ of an event A as the limiting relative frequency of the occurrence of A in an infinite repetition of the experiment. This means that if we repeat the same experiment in such a way that each trial is unaffected by the previous outcomes, then the ratio of the number of occurrences of A to the total number of repetitions of the experiment will converge to $p(A)$. According to this interpretation of probability, the necessary characterizations of the event and probability can be derived quite naturally. For example, we can easily see that the following properties are required of $p(\cdot)$:

$$0 \leq p(A) \leq 1 \qquad\qquad (1)$$

$$p(\Omega) = 1 \qquad\qquad (2)$$

$$p(A_1 \cup A_2) = p(A_1) + p(A_2) \qquad\qquad (3)$$

where $A_1 \cup A_2$ denotes the union of A_1 and A_2 and the intersection of A_1 and A_2 is assumed to be null. To develop a mathematically detailed discussion, the continuity of $p(A)$, i.e., the property that when A_1, A_2, ... successively decreases to the null set ϕ, $p(A_n)$ converges to 0, is further required as a fundamental property of $p(\cdot)$.

Similarly, to represent the necessary collection of events, we may require that F, the whole set of events A, forms a so-called Borel field, i.e., a non-empty set such that, if A is an element of F then A^c ($= \Omega - A$; A's complement) is also an element of F. Further if A_1, A_2, ... are elements of F

then the union $\bigcup_{n=1}^{\infty} A_n$ is also an element of F.

If we consider A, an element of F, as a figure defined in the space Ω, then we may think of $p(A)$ as the measure of the area, or the mass, of the figure, where the total mass over the whole space is normalized to 1.

3.0.2 Random variable

Suppose that event ω occurred in an experiment. When one of ω's characteristics is measured numerically, we represent this by $x(\omega)$. Thus $x(\omega)$ represents a variable whose value is determined by each experiment.

Following the idea developed in the preceding subsection, the event that this variable remains smaller than or equal to a certain value u is represented by

$$\{\omega;\ x(\omega) \leq u\}\ . \tag{1}$$

If the event $\{\omega: x(\omega) \leq u\}$ belongs to F for any value of u, then naturally

$$F(u) = p\{\omega;\ x(\omega) \leq u\} \tag{2}$$

can be defined for any real number u. In such a case, $x(\omega)$ is called a *random variable* and $F(u)(-\infty < u < \infty)$, as a function of u is called the *distribution function* of $x(\omega)$.

When a set of characteristics $x_1, x_2, ..., x_k$ of ω are numerically observed simultaneously, this leads to the concept of the k-dimensional random variable $x(\omega) = \{x_1(\omega), x_2(\omega), ..., x_k(\omega)\}$. In this case the function defined by

$$F(u_1, u_2, ..., u_k) = p\{\omega;\ x_1(\omega) \leq u_1,\ x_2(\omega) \leq u_2,\ ......,\ x_k(\omega) \leq u_k\} \tag{3}$$

is called the *k-dimensional distribution function* of $x(\omega)$. This shows that, if the stochastic structure of the occurrence of the original event ω is completely known, the stochastic structure of the occurrence of any restricted set of characteristics $x_1, x_2, ..., x_k$ is also specified completely.

In the rest of this book, the abbreviated notation x will often be used to represent the random variable $x(\omega)$.

3.0.3 Mean value of a random variable

Suppose that the probabilities for a random variable $x(\omega)$ to take values $a_1, a_2, a_3, ...$ are respectively given by $p_1, p_2, p_3, ...$, where it holds that

$$p_1 + p_2 + p_3 + ... = 1\ .$$

In this case the mean value of x is defined by

$$Ex(\omega) = \sum_{n=1}^{\infty} a_n \, p_n$$

$$= \sum_{n=1}^{\infty} a_n \, p\{x(\omega) = a_n\} . \tag{1}$$

Here E denotes the *expectation*. For an appropriate function of x, $g(x)$, $g(x(\omega))$ defines a random variable and the mean value of $g(x)$ can be obtained as

$$Eg(x(\omega)) = \sum_{n=1}^{\infty} g(a_n) \, p\{x(\omega) = a_n\} . \tag{2}$$

If the distribution function $F(u)$ of $x(\omega)$ is given by

$$F(u) = \int_{-\infty}^{u} f(v) \, dv \tag{3}$$

where $f(v)$ is the *probability density function*, the mean values are obtained as follows:

$$Ex(\omega) = \int_{-\infty}^{\infty} v \, f(v) \, dv \tag{4}$$

and

$$Eg(x(\omega)) = \int_{-\infty}^{\infty} g(v) \, f(v) \, dv . \tag{5}$$

Obviously, the mean values are defined only when the infinite sums or the integrals exist. From the relations (2) and (5) we can see that for arbitrary constants a and b it holds that

$$E(ax(\omega) + b) = aEx(\omega) + b . \tag{6}$$

For a 2-dimensional random variable $(x_1(\omega), x_2(\omega))$ with probability distribution given by

$$p(x_1(\omega) = a_i, x_2(\omega) = b_j) = p_{ij}$$

$$\sum_{i=1}^{\infty} \sum_{j=1}^{\infty} p_{ij} = 1 \tag{7}$$

the mean value of the random variable $g(x_1(\omega), x_2(\omega))$ is given by

$$Eg(x_1(\omega), x_2(\omega)) = \sum_{i=1}^{\infty} \sum_{j=1}^{\infty} g(a_i, b_j) \, p_{ij}$$

$$= \sum_{i=1}^{\infty} \sum_{j=1}^{\infty} g(a_i, b_j) \, p(x_1(\omega) = a_i, x_2(\omega) = b_j). \qquad (8)$$

When the 2-dimensional distribution function $F(u_1, u_2)$ of $x_1(\omega)$ and $x_2(\omega)$ is given by

$$F(u_1, u_2) = \int_{-\infty}^{u_1} \int_{-\infty}^{u_2} f(v_1, v_2) \, dv_1 \, dv_2 \qquad (9)$$

where $f(v_1, v_2)$ is the 2-dimensional probability density function, the mean value of $g(x_1(\omega), x_2(\omega))$ is given by

$$Eg(x_1(\omega), x_2(\omega)) = \int_{-\infty}^{\infty} \int_{-\infty}^{\infty} g(v_1, v_2) \, f(v_1, v_2) \, dv_1 \, dv_2. \qquad (10)$$

The above results can be extended in a natural fashion to the case of a k-dimensional random variable. In any case we can obtain the mean value of a random variable by simply multiplying the possible values of the random variable by the corresponding probabilities (densities), and then summing (integrating) them over all values.

For a random variable $x(\omega)$, define a function $g(x)$ by $g(x)=(x-m)^2$ with $m=Ex(\omega)$. The mean value of $g(x(\omega))$

$$Eg(x(\omega)) = E(x(\omega) - m)^2$$

is called the *variance* of $x(\omega)$, and the positive square root of the variance is called the *standard deviation*.

For a 2-dimensional random variable $(x_1(\omega), x_2(\omega))$

$$E(x_1(\omega) - Ex_1(\omega))(x_2(\omega) - Ex_2(\omega))$$

is called the *covariance* of $x_1(\omega)$ and $x_2(\omega)$. The covariance of $x_1(\omega)$ and $x_2(\omega)$ divided by the product of the corresponding standard deviations

$$\frac{E(x_1(\omega) - Ex_1(\omega))(x_2(\omega) - Ex_2(\omega))}{\sqrt{E(x_1(\omega) - Ex_1(\omega))^2 E(x_2(\omega) - Ex_2(\omega))^2}}$$

is called the *coefficient of correlation* of $x_1(\omega)$ and $x_2(\omega)$.

An important equality concerning the coefficient of correlation is given by

$$\operatorname*{Min}_{a,\,b} E\,(x_2(\omega) - ax_1(\omega) - b)^2 = (1 - r_{1,\,2}^2)\,E\,(x_2(\omega) - Ex_2(\omega))^2 \qquad (11)$$

where $r_{1,2}^2$ denotes the square of the correlation coefficient $r_{1,2}$ of $x_1(\omega)$ and $x_2(\omega)$.

3.0.4   Independence of random variables

Consider a 2-dimensional random variable $(x_1(\omega),\ x_2(\omega))$. If for any arbitrary pair of functions $g_1(x)$ and $g_2(x)$ it holds that

$$Eg_1(x_1(\omega))\,g_2(x_2(\omega)) = Eg_1(x_1(\omega))\,Eg_2(x_2(\omega)) \qquad (1)$$

then by definition $x_1(\omega)$ and $x_2(\omega)$ are *mutually independent*. For the cases treated in 3.0.3 this is equivalent to assuming

$$p(x_1(\omega) = a_i,\ x_2(\omega) = b_j) = p(x_1(\omega) = a_i)\,p(x_2(\omega) = b_j) \qquad (2)$$

or

$$f(v_1,\,v_2) = f_1(v_1)\,f_2(v_2) \qquad (3)$$

where $f_1(v_1)$ and $f_2(v_2)$ denote the probability density functions of $x_1(\omega)$ and $x_2(\omega)$, respectively.

If $x_1(\omega)$ and $x_2(\omega)$ are independent, then, from relations (1) and (6) of 3.0.3, it holds that

$$E(x_1(\omega) - Ex_1(\omega))(x_2(\omega) - Ex_2(\omega))$$
$$= E(x_1(\omega) - Ex_1(\omega))\,E(x_2(\omega) - Ex_2(\omega)) = 0. \qquad (4)$$

Thus the covariance and accordingly the correlation coefficient of $x_1(\omega)$ and $x_2(\omega)$ are equal to 0 in this case.

3.0.5   Gaussian distribution

When the probability density function $f(v)$ of a random variable $x(\omega)$ is given by

$$f(v) = \frac{1}{\sqrt{2\pi}\,\sigma}\exp\left(-\frac{(v-m)^2}{2\sigma^2}\right) \quad (-\infty < v < \infty) \qquad (1)$$

the mean and variance of $x(\omega)$ are equal to m and σ^2, respectively. The probability distribution defined by (1) is called *Gaussian distribution*, or the *normal distribution*, with mean m and variance σ^2.

When ($x_1(\omega), x_2(\omega), ..., x_k(\omega)$) is a k-dimensional random variable, such that for an arbitrary set of coefficients $\{a_1, a_2, ..., a_k\}$ the sum

$$\sum_{i=1}^{k} a_i x_i(\omega) \qquad (2)$$

follows a Gaussian distribution with mean

$$\sum_{i=1}^{k} a_i m_i \qquad (3)$$

and variance

$$\sum_{i=1}^{k} \sum_{j=1}^{k} a_i a_j \sigma_{ij}, \qquad (4)$$

the probability distribution of $(x_1(\omega), x_2(\omega), ..., x_k(\omega))$ is called a k-dimensional Gaussian distribution. Here, m_i of (3) denotes the mean value of $x_i(\omega)$ and σ_{ij} of (4) denotes the covariance of $x_i(\omega)$ and $x_j(\omega)$. The k-dimensional Gaussian distribution is defined uniquely by the mean value vector (m_i; $i=1, 2, ..., k$) and the covariance matrix (σ_{ij}; $i, j=1, 2, ..., k$).

3.0.6 Stochastic processes

To describe the temporal development of a stochastic phenomenon over a period of time, we consider the set of random variables $x(t;\omega)$ which represent the values at the time point t. A set of random variables having a time parameter t is called a *stochastic process*. If we omit ω, a stochastic process defined for discrete time parameter t is represented by

$$\{x(t)\,;\ t=\cdots\cdots,\ -1, 0, 1, 2,\ \cdots\cdots\}\,,$$

and for continuous time parameter t by

$$\{x(t)\,;\ -\infty<t<\infty\}\,.$$

When the distribution of $(x(t_1), x(t_2), ..., x(t_k))$ follows a k-dimensional Gaussian distribution for any arbitrarily selected set of time points $(t_1, t_2, ..., t_k)$(k is also arbitrary), the stochastic process $\{x(t)\}$ is called a *Gaussian process.*

When the structure of a stochastic process does not depend on the choice of the origin of time and is temporally homogeneous i.e., for an arbitrary choice of k, t_1, t_2, ..., t_k, and t, the probability distribution of $x(t_1)$, $x(t_2)$, ..., $x(t_k)$ and that of $x(t_1+t)$, $x(t_2+t)$, ..., $x(t_k+t)$ are identical, $\{x(t)\}$ is called a *stationary stochastic process.* Sometimes, instead of the equality of the distributions, only the equality of the means and covariances is assumed. This type of stochastic process is called a *weakly stationary stochastic process* or a *stationary stochastic process in the wide sense.* However, this definition will not be used in the present book. The definition of the stationary stochastic process can naturally be extended to the case where $x(t)$ is a p-dimensional vector. In this case it is called a *p-dimensional stationary stochastic process.*

Some stationary stochastic processes are called *ergodic.* Ergodicity means that the behavior of the process becomes, with the lapse of time, independent of the behavior in the remote past. For example, it is known that a stationary Gaussian process with continuous time parameter is ergodic, if the covariance $Ex(s+t)\,x(t)$ is continuous with respect to the time-difference s, and at the same time it converges to zero as s is increased indefinitely. To assume a stochastic process to be ergodic is equivalent to assuming the structure of the process is to be determined perfectly from an infinitely long record of a single observation of the process.

Completing the brief introduction of some basic probabilistic or statistical concepts we now turn to the discussion of other concepts that are more directly related to the handling of real data.

3.1 Spectral Analysis of Stationary Time Series

For the analysis of a phenomenon showing a stationary irregular fluctuation, it is often effective to observe the result of decomposition of the fluctuation into periodic components. In this section we will explain the basic principle and practical use of this procedure. Concrete examples of application of the concepts and methods described in this chapter will be given in Chapter 4. Therefore, it is advisable for the reader to go back and forth between Chapter 4 and the present chapter, following the progress of his understanding of the subject.

A series of observations of a randomly fluctuating phenomenon generated by a time-invariant mechanism is called a *stationary time series.* Here in this book, we will restrict our attention to those time series which, when expressed mathematically, can be taken as realizations of an ergodic stationary stochastic process. This is equivalent to assuming the series has the following property:

For any function of the time series, consider the series of values obtained by successively shifting the origin of time of the time series by unit time in the difinition of the original function. Then the average of the series thus obtained is identical, with probability 1, to the probabilistic mean value of the original function, which can be defined as the average of the values of the function obtained by replacing the original time series by other independent realizations from the process.

This means that we are assuming that it is possible to infer about all the possible outcomes of the experiment by observing a single infinite length realization of the experiment. We may assume this for a phenomenon generated by a temporally invariant system whose memory dies out with the lapse of time, i.e., its motion eventually becomes independent of the past.

Hereafter, we consider a series of data observed at each integral multiple of a constant time interval. The effect of this time-sampling from a time series with continuous time parameter will be discussed in 3.4. We denote the observation at time point t by $x(t)$. If we denote the sampling interval by Δt the given data is represented by

$$\{x(s\Delta t)\,;\ s=1, 2, \cdots\cdots, N\}\,.$$

In the following, as long as there is no danger of confusion, we will consider Δt to be equal to 1 and write $x(s)$ in place of $x(s\Delta t)$. The symbol s should not be confused with the argument of the Laplace transform discussed in Chapter 2. When the physical meaning of Δt gains importance we will use Δt explicitly.

3.1.1 Autocorrelation function and power spectrum

Suppose we have a stationary time series $\{x(s)\,;\ s=1, 2, ..., N\}$, showing irregular fluctuations. In order to describe the periodic characteristic of its motion, it is possible to think of a method of decomposing the motion into its frequency components. This leads to an expression of the distribution of the components over the frequency axis, just like the spectrum or decomposition of light obtained by passing it through a prism. Examples of this type of decomposition are given in 4.1

The *autocovariance function* $R_{xx}(l)$ $(l=0, \pm1, \pm2, \cdots\cdots)$ of $x(s)$ is defined by

$$R_{xx}(l)=\lim_{N\to\infty}\frac{1}{N}\sum_{s=1}^{N}(x(s+l)-m_x)(x(s)-m_x) \tag{1}$$

where

$$m_x = \lim_{N \to \infty} \frac{1}{N} \sum_{s=1}^{N} x(s). \qquad (2)$$

However, when Δt is considered, $R_{xx}(l)$ must be written as $R_{xx}(l\Delta t)$. We assume that

$$\sum_{l=-\infty}^{\infty} |R_{xx}(l)| < \infty.$$

Under this assumption the *power spectral density function* $q_{xx}(g)$ of $x(s)$ is given by the Fourier transformation of $\{R_{xx}(l)\}$, i.e., it is given by

$$q_{xx}(g) = \sum_{l=-\infty}^{\infty} \exp(-i2\pi gl) R_{xx}(l) \quad (i^2 = -1), \qquad (3)$$

where g denotes the frequency, and also

$$R_{xx}(l) = \int_{-\frac{1}{2}}^{\frac{1}{2}} \exp(i2\pi gl) q_{xx}(g) \, dg. \qquad (4)$$

$q_{xx}(g)$ gives the power spectral density of $x(s)$ at frequency $g(\text{cycle}/\Delta t)$ where Δt is the sampling interval. If we denote the power spectral density of $x(s)$ at frequency $f(\text{cycle}/\text{unit time})$ by $p_{xx}(f; \Delta t)$ then by transforming the variable g in (4) into $f = g/\Delta t$ we get, for $-1/2\Delta t \leq f \leq 1/2\Delta t$,

$$R_{xx}(l) = \int_{-\frac{1}{2\Delta t}}^{\frac{1}{2\Delta t}} \exp(i2\pi f l \Delta t) p_{xx}(f; \Delta t) \, df. \qquad (5)$$

Thus we get

$$\Delta t \, q_{xx}(g) = p_{xx}\left(\frac{g}{\Delta t}; \Delta t\right), \qquad (6)$$

and by (3) we get

$$p_{xx}(f; \Delta t) = \Delta t \sum_{l=-\infty}^{\infty} \exp(-i2\pi f l \Delta t) R_{xx}(l). \qquad (7)$$

When the original phenomenon is continuous in time t, we consider only such cases where the relations obtained by making Δt approach 0 in (5) and (6) hold, i.e., we assume the relations

$$p_{xx}(f) = \int_{-\infty}^{\infty} \exp\left(-i\,2\pi f\tau\right) R_{xx}(\tau)\,d\tau \quad (-\infty < f < \infty) \tag{8}$$

and

$$R_{xx}(\tau) = \int_{-\infty}^{\infty} \exp\left(i\,2\pi f\tau\right) p_{xx}(f)\,df. \tag{9}$$

As in the present discussion the existence of the necessary Fourier transforms will be assumed throughout the rest of this book. It should be noted here that the covariance function $R_{xx}(\tau)$ is defined by

$$R_{xx}(\tau) = \lim_{T \to \infty} \frac{1}{T} \int_{0}^{T} (x(t+\tau) - m_x)(x(t) - m_x)\,dt, \tag{10}$$

where

$$m_x = \lim_{T \to \infty} \frac{1}{T} \int_{0}^{T} x(t)\,dt. \tag{11}$$

When Δt is sufficiently small, $p_{xx}(f; \Delta t)$ gives practically sufficient approximation of $p_{xx}(f)$. We will discuss this point later in 3.4. Here we will only assume that Δt is small enough so that an integral with respect to time of form (8) can be approximated sufficiently accurately by the corresponding sum of form (7).

The most fundamental relation concerning the power spectral density function $p_{xx}(f)$ is given when the original $x(t)$ undergoes a linear transformation defined by

$$y(t) = \int_{-\infty}^{\infty} x(t-\tau)\,h(\tau)\,d\tau. \tag{12}$$

The power spectral density function $p_{yy}(f)$ of $y(t)$ is then given by

$$p_{yy}(f) = |A(f)|^2\, p_{xx}(f) \quad (-\infty < f < \infty), \tag{13}$$

where

$$A(f) = \int_{-\infty}^{\infty} \exp\left(-i\,2\pi f\tau\right) h(\tau)\,d\tau. \tag{14}$$

$A(f)$ is called the *frequency response function* of the time invariant linear system which realized the transformation. The corresponding function $h(t)$ is called the *impulse response function* of the system. For a time series with discrete time we assume that $\Delta t = 1$ and (12), (13) and (14) are replaced respectively by

$$y(s) = \sum_{m=-\infty}^{\infty} a(m) \, x(s-m),\tag{15}$$

$$p_{yy}(f) = |A(f)|^2 \, p_{xx}(f) \quad \left(-\frac{1}{2} \leq f \leq \frac{1}{2}\right),\tag{16}$$

and

$$A(f) = \sum_{m=-\infty}^{\infty} \exp\left(-i\,2\pi f m\right) a(m) \quad \left(-\frac{1}{2} \leq f \leq \frac{1}{2}\right).\tag{17}$$

In this case, $A(f)$ of (17) is called the frequency response function of the system and $\{a(m)\}$ the impulse response function. Since the correspondence between the time-continuous and discrete cases is obvious we will develop our discussion without discriminating between the two cases. Formula (13) give the fundamental relation that explains the influence of a linear system on the power spectral density of the input. Once the power spectral densities $p_{xx}(f)$ and $p_{yy}(f)$ of the input $x(t)$ and output $y(t)$ of the system are given, $|A(f)|^2$ can be calculated and thus the effect of the system can be grasped quantitatively.

Besides its importance due to the above mentioned analytic property, the power spectral density function has yet another important characteristic when it is viewed from the standpoint of information processing. This is the ease of understanding the output. Mathematically the autocovariance function $R_{xx}(t)$ and the power spectral density function $p_{xx}(f)$ are equivalent. However, for practical applications, there is a great difference between the two. The spectral density function $p_{xx}(f)$ has two advantages over $R_{xx}(t)$. Its physical meaning can easily be understood with the aid of relations such as (13). Also it is easily perceived by the human eye. The estimation procedure for the power spectral density using a record of observational data will be discussed later. The practical utility of the power spectral density function, or the power spectrum, can be recognized quickly by simply looking at examples in 4.1.

3.1.2   Cross-covariance function and cross-spectrum
    Here we consider a 2-dimensional stationary time series,

$$\{x(s), y(s); s = 1, 2, \ldots, N\}.$$

The *cross-covariance function*. or cross-covariance sequence, $R_{yx}(l)$ ($l=0, \pm 1, \pm 2, \ldots$) of $x(s)$ and $y(s)$ is defined by

$$R_{yx}(l) = \lim_{N \to \infty} \frac{1}{N} \sum_{s=1}^{N} (y(s+l) - m_y)(x(s) - m_x), \qquad (1)$$

where

$$m_x = \lim_{N \to \infty} \frac{1}{N} \sum_{s=1}^{N} x(s), \qquad m_y = \lim_{N \to \infty} \frac{1}{N} \sum_{s=1}^{N} y(s). \qquad (2)$$

Correspondingly, the *cross-spectral density function* of $x(s)$ and $y(s)$ is defined by

$$p_{yx}(f) = \sum_{l=-\infty}^{\infty} \exp(-i 2\pi f l) R_{yx}(l) \quad \left(-\frac{1}{2} \le f \le \frac{1}{2}\right), \qquad (3)$$

and satisfies the following relation

$$R_{yx}(l) = \int_{-\frac{1}{2}}^{\frac{1}{2}} \exp(i 2\pi f l) \, p_{yx}(f) \, df. \qquad (4)$$

When we consider a continuous time parameter t, we have only to replace the summation in (3) by the corresponding integral and define the range of f by $-\infty < f < \infty$. Necessary modifications of (1), (2) and (4) are obvious. Corresponding to the fundamental characteristic of the power spectral density, given by (13) or (16) of the preceding subsection 3.1.1 where y is obtained from x by the transformation (12) or (15) of 3.1.1, it holds that

$$p_{yx}(f) = A(f) \, p_{xx}(f), \qquad (5)$$

provided that the range of f is appropriately chosen in each case. This equality (5) forms the basis for the determination of the frequency response function of a time invariant linear system by statistical methods.

When the time series $\{x(s), z(s); s=1, 2, \ldots, N\}$ represents the record of observations of an actual stationary system operating with a stochastic input x and output z, the relation between the output z and x is not given by a strictly linear relation as in (12) or (15) of 3.1.1 for the input x and output y. Generally, the observed output z can be written as a sum of two components,

$$z(s) = y(s) + n(s), \qquad (6)$$

where y is strictly linearly related with x and n is linearly unrelated, i.e., $n(s)$ satisfies the following relation, which is a special case of (5) with $A(f)$ equal to 0;

$$p_{nx}(f)=0. \tag{7}$$

This is also equivalent to the condition

$$R_{nx}(l)=0 \quad (l=0, \pm 1, \pm 2, \cdots\cdots). \tag{8}$$

The time series $n(s)$ can be considered as a mixture of the noise that has nothing to do with the input $x(s)$ and the effect of the nonlinearity of the system. In this case, for the autocovariance function, it holds that

$$R_{zz}(l)=R_{yy}(l)+R_{nn}(l), \tag{9}$$

and correspondingly, for the power spectral density function,

$$p_{zz}(f)=p_{yy}(f)+p_{nn}(f). \tag{10}$$

At each frequency we consider the ratio of the power of y which is in linear relation with the input x to the power of z

$$\gamma^{2}(f)=\frac{p_{yy}(f)}{p_{zz}(f)} \tag{11}$$

which is called the *coherency* of z and x at f. When $\gamma^{2}(f)$ is considered as a function of f, it is called the *coherency function* of z and x. From (7) it hols that

$$p_{zx}(f)=p_{yx}(f)+p_{nx}(f)=p_{yx}(f). \tag{12}$$

Thus, for the estimation of the frequency response function $A(f)$ of a linear system, we can replace $p_{yx}(f)$ by $p_{zx}(f)$ in (5) and use the relation

$$A(f)=\frac{p_{zx}(f)}{p_{xx}(f)}. \tag{13}$$

This shows that theoretically it does not matter whether z is contaminated by noise $n(s)$ or not. This is the biggest merit of estimating $A(f)$ by statistical methods.

If we define $A(f)$ by (13), $p_{yy}(f)$ of (10) is given by

$$p_{yy}(f) = |A(f)|^2 p_{xx}(f) \tag{14}$$

and $\gamma^2(f)$ can be represented as

$$\gamma^2(f) = \frac{|p_{zx}(f)|^2}{p_{zz}(f) \, p_{xx}(f)} . \tag{15}$$

Therefore, we can see that $\gamma^2(f)$ is a symmetrical quantity with respect to z and x. Intuitively speaking, $\gamma^2(f)$ denotes the square of the correlation coefficient between the frequency components of z and x at frequency f. It is obvious that this concept of coherency will play an important role in the analysis of a statistical system. When estimating a frequency response function by a statistical method, this quantity has a decisive effect on the accuracy of the estimate. For an example of the estimation of coherencies, readers are referred to 4.1.

It must be mentioned here that, when estimating a frequency response function of a linear system by (13), the input x must be measured without observation error. Unlike the noise $n(s)$ in the system, the effect of this measurement error cannot be eliminated by the present method. In the following discussion, it will always be assumed that the measurement error is small enough to be ignored. This is a rather strong assumption and its validity must be checked carefully when the method is put into practical use.

3.1.3 Spectral analysis of a multiple input system

It is possible to extend the discussion of the previous section to the case of a stationary system operating with a multiple input. For this purpose we use the following simple relationf for $y = u + v$:

$$R_{yx}(l) = R_{u+v,x}(l) = R_{ux}(l) + R_{vx}(l) . \tag{1}$$

Consider a stationary system generating the output x_0 from the inputs x_1, x_2, ..., x_k. Then x_0 can be expressed as a sum of two parts, one in a linear relation with $x_1, x_2, ..., x_k$ and the other without any linear relation with the inputs. We consider that the part linearly related with the input can be expressed as a sum of the linear transforms of the inputs x_l with the corresponding frequency response functions $A_l(f)$. In this case, the following relation holds:

$$\sum_{l=1}^{k} A_l(f) \, p_{lm}(f) = p_{0m}(f) \quad (m = 1, 2, \cdots\cdots, k) , \tag{2}$$

where $p_{lm}(f)$ denotes the cross-spectral density function of x_l and x_m, or in the case $l=m$, the power spectral density function of x_l.

When the $p_{lm}(f)$'s are given, (2) defines a first order simultaneous equation of complex variables with $A_l(f)(l=1, 2, ..., k)$ as the unknowns. If the equation has a unique solution, the frequency response functions are uniquely determined. In this case, the power spectral density of the part linearly related to the inputs is given by

$$p_{pp}(f) = \sum_{l=1}^{k} \sum_{m=1}^{k} A_l(f) \overline{A_m(f)} \, p_{lm}(f)$$

$$= \sum_{m=1}^{k} p_{0m}(f) \overline{A_m(f)} \,, \qquad (3)$$

where — denotes complex conjugate. Using this relation, we define, to correspond to $\gamma^2(f)$,

$$\gamma^2_{0,\,12\,\cdots\cdots\,k}(f) = \frac{p_{pp}(f)}{p_{00}(f)}$$

and call this the *multiple coherency* of x_0 and $x_1, x_2, ..., x_k$. There is another important concept of partial coherency. However, since its definition is rather complicated, we will not give its theoretical definition here. Only a definition of its estimate will be given later.

3.1.4   Statistical estimation of spectra.

Let us first examine the effect of sampling data at constant time intervals. For this purpose we have only to examine the effect on the fundamental frequency component $\exp(i2\pi ft)$. This function has frequency f and gives the ordinate of a point traveling with fixed angular speed $2\pi f$ radians per unit time on the unit circle. To understand the movement of observations of this point at times $t=s\Delta t(s=0, \pm 1, \pm 2, ...; \Delta t>0)$ imagine the rotation of a wheel of a station wagon in a western movie. If Δt is the interval between the consecutive shutter openings any rotation of the wheel completed during an interval of Δt is ignored and only the residual movement that is within the limits $\pm \pi$ radians is recognized. This situation is represented by the following relation:

$$\exp\left(i\,2\pi\left(f+\frac{k}{\Delta t}\right)s\,\Delta t\right) = \exp\left(i\,2\pi f s\,\Delta t\right), \qquad (1)$$

where the range of f is limited to $-1/(2\Delta t) \leq f \leq 1/(2\Delta t)$. This relation clearly shows that by this observation procedure the frequency components with frequencies differing from f by integral multiples of $1/\Delta t$ are recognized as having one and the same frequency f. Similarly, from (9) of 3.1.1 we get

$$R_{xx}(l\,\Delta t) = \int_{-\infty}^{\infty} \exp\left(i\,2\pi f l\,\Delta t\right)\, p_{xx}(f)\, df$$

$$= \int_{-\frac{1}{2\Delta t}}^{\frac{1}{2\Delta t}} \exp\left(i\,2\pi f l\,\Delta t\right)\, p_{xx}(f;\,\Delta t)\, df\,, \qquad (2)$$

where

$$p_{xx}(f;\,\Delta t) = \sum_{k=-\infty}^{\infty} p_{xx}\left(f+\frac{k}{\Delta t}\right)\quad\left(-\frac{1}{2\Delta t}\leq f\leq\frac{1}{2\Delta t}\right). \qquad (3)$$

The formula (2) is the same as (5) of 3.1.1, except that here Δt is restored and $R_{xx}(l)$ is replaced by $R_{xx}(l\Delta t)$.

When only $x(s\Delta t)$ are observed, all we can obtain are the values of $R_{xx}(l\Delta t)(l=0, \pm 1, \ldots\ldots)$ and $p_{xx}(f;\Delta t)$ given by (2) and (3). This shows that $p_{xx}(f;\Delta t)$ given by (5) of 3.1.1 is related to the original $p_{xx}(f)$ by the above equality (3), i.e., the total power of the components having differences in frequency from f, $0\leq f\leq 1/(2\Delta t)$, by integral multiples of $1/\Delta t$ accumulates at frequency f. Since we consider only real-valued $x(t)$ it holds that

$$R_{xx}(\tau) = R_{xx}(-\tau)$$

and thus we have

$$p_{xx}(-f) = p_{xx}(f).$$

From this relation, to obtain the value of $p_{xx}(f;\Delta t)$ within the range of $0\leq f\leq 1/(2\Delta t)$, we may imagine a procedure where the figure of $p_{xx}(f)$ is cut out of paper and folded at $f=k/(2\Delta t)(k=1, 2, \ldots)$, successively to the left and right and the values of the piled up portions of the spectrum are added to get $p_{xx}(f;\Delta t)$. Following this idea, we call $1/(2\Delta t)$ the *folding frequency* of the sampling procedure with sampling interval Δt. The phenomenon that components with frequencies differing by $1/\Delta t$ appear identical is called *aliasing* and $p_{xx}(f;\Delta t)$ is sometimes called the *aliased spectrum*. A similar relation holds for the cross-spectral density function. For the determination of the sampling interval Δt, the effect of aliasing must be considered carefully.

Estimation of the Power Spectral Density Function (See Program 5.1.1 or 5.1.3, and Program 5.2.1)

Here we discuss the Blackman-Tukey type procedure [2]. The new method proposed by one of the present authors will be discussed in the next section.

[Constants for the preparation of original data]

Δt: sampling interval, to be determined to be *as large as* possible so long as the effect of aliasing due to the existence of power above the frequency $1/2\Delta t$ can be ignored.

W: width of the window, defines the width of the frequency band over which the smoothing operation is performed and controls the resolvability of the resulting estimate.

N: number of data points, should be as large as possible and satisfy the relation

$$N\Delta t > 10\frac{1}{W} \,.$$

[Estimation procedure]

Assume that a data set $\{x(s\Delta t); s=1, 2, ..., N\}$ is given.

1) Eliminate dc(direct current) component or the average value of the data to define $x(s)$ by

$$x(s) = x(s\,\Delta t) - \bar{x} \ (s=1, 2, ..., N), \quad \text{where} \quad \bar{x} = \frac{1}{N}\sum_{s=1}^{N} x(s\Delta t) \,.$$

2) Determine an integer h that satisfies the approximate equality

$$\frac{1}{h\Delta t} \approx W \,.$$

3) Compute (sample) autocovariances

$$C_{xx}(l) = \frac{1}{N}\sum_{s=1}^{N-l} x(s+l)\, x(s) \quad (l=0, 1, \cdots\cdots, h) \,.$$

4) Obtain the raw Fourier transforms of the truncated autocovariances by

$$P(r) = \Delta t \left\{ C_{xx}(0) + 2 \sum_{l=1}^{h-1} \cos\left(2\pi \frac{r}{2h} l\right) C_{xx}(l) + (-1)^r C_{xx}(h) \right\}$$

$$(r = 0, 1, \cdots\cdots, h).$$

5) Obtain smoothed Fourier transforms by the windowing procedures

$$P_i(r) = \sum_{n=-2}^{2} a_i(n) \, P(r-n) \quad (i=1, 2; \, r=0, 1, \cdots\cdots, h),$$

where $a_i(n)$'s are given by*

$$a_1(0) = 0.5 \qquad a_1(1) = a_1(-1) = 0.25 \qquad a_1(2) = a_1(-2) = 0,$$

or

$$a_2(0) = 0.625 \qquad a_2(1) = a_2(-1) = 0.25 \qquad a_2(2) = a_2(-2) = -0.0625,$$

and put

$$P(-n) = P(n), \qquad P(h+n) = P(h-n).$$

6) $P_1(r)$ gives the estimate of

$$p_{xx}\left(\frac{r}{2h\Delta t}\right).$$

7) When there are many r's for which the following inequality holds

$$\left| \frac{P_2(r) - P_1(r)}{P_1(r)} \right| > 0.43 \sqrt{\frac{h}{M}},$$

and if they are within the frequency range of interest, double the value of h and recalculate the estimates. When the difference between $P_2(r)$ and $P_1(r)$ is large, that will mean that the curvature of $p_{xx}(f)$ at that frequency was so large that the smoothing operation using the Hanning window produced significant bias [3].

Remarks:

1. If the windows defined in 5) are used $C_{xx}(h)$ may be equated to 0

---

*The smoothing kernel defined by the coefficients $\{a_1(n)\}$ is know as the Hanning window [2]. The window defined by $\{a_2(n)\}$ was introduced by the first author of the present book.

without computation.

2. The above step 4) is the numerical Fourier (cosine) transformation of the sequence $\{\Delta t C_{xx}(0), 2\Delta t C_{xx}(l)(l=1, 2, \ldots\ldots, h-1), \Delta t C_{xx}(h)\}$ and an efficient computational procedure is realized by Goertzel's method. See programs in 5.2.

3. If we define $D_i(l)$ by

$$D_i(l) = \sum_{n=-2}^{2} a_i(n) \cos\left(2\pi \frac{n}{2h} l\right) \quad (|l| \leq h)$$

and use $D_i(l)C_{xx}(l)$ ($l=0, 1, \ldots, h$) in place of $C_{xx}(l)$ in step 4, the Fourier transform is equal to $P_i(r)$ and step 5) becomes unnecessary. In this case, by using $\cos(2\pi f)$ instead of $\cos(2\pi rl/2h)$ the estimate of $p_{xx}(f)$ at an arbitrary frequency f can be obtained.

4. When N and h are large, the covariance computation of step 3) can be made faster by using the FFT (Fast Fourier Transform).

5. The standard deviation of the estimate of $p_{xx}(f)$ may be considered to be about the size of $(h/N)^{1/2} \times p_{xx}(f)$.

Estimation of the Frequency Response Function and Coherency [4, 5, 6] (See Programs 5.1.2, 5.1.3, 5.2.2, 5.2.3, 5.2.4.)

Let x_0 denote the output and $x_1, x_2, \ldots\ldots, x_k$ the inputs of the system. Observations are given by $x_i(s\Delta t)(i=0, 1, \ldots\ldots, k; s=1, 2, \ldots\ldots, N)$.

[Constants for the preparation of original data]

Δt: sampling interval, to be determined as in the case of power spectrum estimation.

h: lag length, to be chosen in such a way that the effect of averaging over a frequency band of width $1/h\Delta t$ would not significantly change the shape of the frequency response function. A final decision should be made on the basis of the results of several trials.

N: length of data, which should be large enough to satisfy the inequality

$$N \geq 5h.$$

The larger N becomes, the more accurate the estimates become.

[Estimation procedure]

1) Subtract the dc component, or mean value, and denote the resulting data by $\{x_0(s), x_1(s), \ldots, x_k(s); s=1, 2, \ldots, N\}$.

2) For j, $l=0, 1, \ldots, k$, compute the auto and cross covariances by

$$C_{jl}(m) = \frac{1}{N} \sum_{s=1}^{N-m} x_j(m+s) \, x_l(s) \quad (h \geq m \geq 0)$$

and replace $C_{jl}(h)$ by $(1/2) \, C_{jl}(h)$.

3) For $j=0, 1, ..., k$; $l=j, j+1, ..., k$, and $r=0, 1, ..., h$, calculate

$$Q_{jl}(r) = \sum_{m=-h}^{h} \exp\left(-i \, 2\pi \frac{r}{2h} m\right) C_{jl}(m) \, \varDelta t \, ,$$

where

$$C_{jl}(-m) = C_{lj}(m) \quad (m=1, 2, \cdots\cdots, h) \, .$$

4) For $j=0, 1, ..., k$; $l=j, j+1, ..., k$, compute

$$P_{jl}(r; i) = \sum_{n=-2}^{2} a_i(n) \, Q_{jl}(r-n) \, .$$

The coefficients $a_i(n)$ are the same as those used for power spectral density estimation. Note the relation

$$Q_{jl}(-n) = \overline{Q_{jl}(n)}, \quad Q_{jl}(h+n) = \overline{Q_{jl}(h-n)} \, .$$

5) $P_{jl}(r) = P_{jl}(r; 1)$ gives the estimate of the cross spectral density $p_{jl}(r/(2h \varDelta t))$.

6) Using the $P_{jl}(r)$'s, set up the following matrix:

	1	2		k	$k+1$	$k+2$	$k+3$		$2k+1$
1	$P_{11}(r)$	$P_{12}(r)$	······	$P_{1k}(r)$	$P_{10}(r)$	1	0	······	0
2	$P_{21}(r)$	$P_{22}(r)$	······	$P_{2k}(r)$	$P_{20}(r)$	0	1	······	0
	$\vdots$	$\vdots$		$\vdots$	$\vdots$	$\vdots$	$\vdots$		$\vdots$
k	$P_{k1}(r)$	$P_{k2}(r)$	······	$P_{kk}(r)$	$P_{k0}(r)$	0	0	······	1
0	$P_{01}(r)$	$P_{02}(r)$	······	$P_{0k}(r)$	$P_{00}(r)$	0	0	······	0

where it is assumed that $P_{jl}(r) = \overline{P_{jl}(r)}$ ($j=0, 1, ..., k$; $l=j+1, ..., k-1$).
In the columns following the $(k+1)$-th, only the elements of the 1st, 2nd, ..., k-th rows, respectively, are equal to 1 and the rest are zeros.

7) To the above matrix, apply the following operations to the rows.

a) Divide the 1st row by $P_{11}(r)$.

b) Multiply the 1st row by $P_{jl}(r)$ and subtract it from the j-th row ($j=2, 4$,

..., k, 0).

By this operation, in the 1st column the element in the 1st row is transformed into 1 and the remaining elements are transformed into 0.

c) After the above operation divide the 2nd row by the (2,2)-th element of the resulting matrix.

d) For $j=1, 3, ..., k, 0$ apply an operation similar to that of b) to the j-th row so that the (2,2)-th element of the resulting matrix becomes 1 and other elements in the 2nd column become zero.

e) Repeat similar operations for all rows up to the k-th. Then the result is:

$$\begin{bmatrix} 1 & 0 & \cdots\cdots & 0 & \alpha_1 & r_{11} & r_{12} & \cdots\cdots & r_{1k} \\ 0 & 1 & \cdots\cdots & 0 & \alpha_2 & r_{21} & r_{22} & \cdots\cdots & r_{2k} \\ \vdots & \vdots & & \vdots & \vdots & \vdots & \vdots & & \vdots \\ 0 & 0 & \cdots\cdots & 1 & \alpha_k & r_{k1} & r_{k2} & \cdots\cdots & r_{kk} \\ 0 & 0 & \cdots\cdots & 0 & \varepsilon & \beta_1 & \beta_2 & \cdots\cdots & \beta_k \end{bmatrix}.$$

8) The estimate of $A_j(f)$ at $f_r=r/(2h\, \Delta t)$ is given by

$$\hat{A}_j(r) = -\beta_j \quad (=\bar{\alpha}_j).$$

(See remark 2 of 3.2.2 on the fitting of multivariate autoregressive models.)

9) The estimate of the multiple coherency between x_0 and $x_1, x_2, ..., x_k$ at frequency f_r is given by

$$\hat{r}^2_{0,12\cdots\cdots k}(r) = 1 - \frac{\varepsilon}{P_{00}(r)}.$$

10) The estimate of the partial coherency between x_0 and x_j at f_r is given by

$$\hat{r}^2_{0j,12\cdots\cdots\hat{j}\cdots\cdots k}(r) = \frac{|\alpha_j|^2}{|\alpha_j|^2 + \varepsilon r_{jj}}.$$

11) A measure of the relative error $R_j(r, \delta)$ that approximately satisfies the relation

$$\text{Prob}\,(|\hat{A}_j(r) - A_j(f_r)| \leq |\hat{A}_j(r)| \cdot R_j(r, \delta)) = \delta$$

is given by

$$R_j(r, \delta) = \sqrt{\frac{1}{d-k} \left(\frac{1}{\hat{\gamma}^2_{0j,12\cdots\hat{j}\cdots k}} - 1 \right) F(2; 2(d-k), \delta)}$$

where $F(2; 2(d-k), \delta)$ is defined by the relation

$$\text{Prob } (F(2; 2(d-k), \delta) \geq F^2_{2(d-k)}) = \delta$$

where $F_{2(d-k)}^2$ denotes a random variable following an F-distribution with degrees of freedom 2 and $2(d-k)$ where d is an integer closest to

$$d = \frac{N}{h} \frac{1}{2 \sum\limits_{n=-1}^{1} (a_1(n))^2} .$$

The value of $R_j(r, \delta)$ with $\delta = 0.95$ can be used as a standard for the relative error of $A_j(r)$. Obviously the number of lags h should be such that the inequality $d > k$ is satisfied.

12) Repeat the same computation with $P_{jl}(r) = P_{jl}(r;2)$.

If the value of the newly obtained $A_j(r)$ differs significantly from the previous one, double the value of h and recalculate.

Remarks:

1. The *partial coherency* $\gamma^2_{0j,12\cdots j\cdots k}(f)$ is equal to the coherency between x_0 and x_l after subtracting the portions linearly related with other input variables in $x_1, x_2, \ldots, x_k$.

2. As k becomes large, $R_j(r, \delta)$, giving the magnitude of the relative error of the estimate, often tends to be large. To avoid this, the number d and accordingly the data length N must be increased.

3. To obtain a meaningful result by applying the above method to the record of inputs and output of a system operating under randomly fluctuating inputs, the feedback from the output to the inputs must be small enough to be ignored. An exception is the case where the output does not contain any noise that is not linearly related with the inputs. For the details of this point, see 3.3.

3.2 Autoregressive Model Fitting

In 3.1 the discussion was developed from the point of view of spectrum analysis which decomposes irregular fluctuations into periodic components. Here the characteristics of a system were represented in the form of a frequency response function and statistical procedures were developed for the estimation of the spectral density function, frequency response function, etc. The power spectrum of a stationary time series was the starting point of Wiener's prediction theory. Once the power spectral density function was given, the best linear prediction in the sense of mean squared error could systematically be realized (see [1] of Chapter 1).

Conversely, as will be seen shortly, if the formula that defines the best linear predictor is given, this directly yields the form of the power spectral density. Therefore, which of these formulations should be adopted as the starting point of discussion depends on practical convenience.

Attempts to realize the best linear prediction on the basis of the observed data $\{x(s); s=1, 2, ..., N\}$ have long been known, and the subject has been treated by the method of last squares. However, when the calculation was based on a data set of finite length, there was a difficulty in determining to what extent the values of the past observations should be used in realizing the prediction. This problem remained unsolved, as also did the problem of determining the number of lags h needed for the estimates of the auto and cross covariance functions used in the spectrum estimation.

Recently one very practical solution to this problem was proposed by one of the present authors and this prepared the starting point for the development of a basic statistical procedure for the practical application of the controller design to be discussed in this book [7]. The significance of this solution has been discussed also from the standpoint of information theory (see [2] of Chapter 1).

3.2.1   Univariate autoregressive model

By subtracting from a stationary time series $x(s)$ the portion that can be expressed as a linear combination of M past values $x(s-1), ..., x(s-M)$, we get the series of residuals

$$\varepsilon(s) = x(s) - \sum_{m=1}^{M} a(m)\, x(s-m). \tag{1}$$

Assume here that an infinitely long series of data is available and that the dc component of $x(s)$ given by

$$m_x = \lim_{N \to \infty} \frac{1}{N} \sum_{s=1}^{N} x(s)$$

is zero. In this case we have

$$\overline{\varepsilon^2(s)} = \lim_{N \to \infty} \frac{1}{N} \sum_{s=1}^{N} \varepsilon^2(s)$$

$$= R_{xx}(0) - 2 \sum_{m=1}^{M} a(m) \, R_{xx}(m) + \sum_{l=1}^{M} \sum_{m=1}^{M} a(l) \, a(m) \, R_{xx}(m-l) \qquad (2)$$

and if only the autovariances $R_{xx}(0)$, $R_{xx}(1)$, ..., $R_{xx}(M)$ are given, the coefficients $a(1)$, $a(2)$, ..., $a(M)$ of the *best linear predictor* that minimizes the mean squared error $\overline{\varepsilon^2(s)}$ can be determined.

For this set of coefficients, the corresponding prediction error $\varepsilon(s)$ is uncorrelated with the past M values $x(s-1)$, ..., $x(s-M)$. If the correlations among the original series $x(s)$ are such that the correlation diminishes to zero as the separation in time of two observations is increased indefinitely, the covariances between $\varepsilon(s)$ and the past values $x(s-M-1)$, $x(s-M-2)$, ..., will also become small when M is increased. Thus, if the $a(m)$'s define the best linear predictor for a sufficiently large M, the corresponding $\varepsilon(s)$ may practically be considered to be uncorrelated with all the past values of $x(s)$. In such a case, since $\varepsilon(s-m)$ with $m>1$ is given as a linear combination of the past values $x(s-m)$, $x(s-m-1)$, ..., the covariance between $\varepsilon(s)$ and $\varepsilon(s-m)$ will be zero. Namely $\varepsilon(s)$ has an autocovariance function defined by

$$R_{\varepsilon\varepsilon}(m) = \delta_{m,0} \, \sigma^2 \quad (\sigma^2 > 0) \qquad (3)$$

where $\delta_{m,0} = 1$ for $m=0$, and $=0$, otherwise.

From (3) of 3.1.1 the power spectral density of $\varepsilon(s)$ is given by

$$q_{\varepsilon\varepsilon}(g) = \sigma^2 \quad \left(-\frac{1}{2} \leq g \leq \frac{1}{2} \right) \qquad (4)$$

i.e., $\varepsilon(s)$ has a flat spectrum. As this resembles the spectrum of the energy distribution of white light, we often call such an $\varepsilon(s)$ a discrete time parameter *white noise*. $\varepsilon(s)$ is also called the *innovation* of $x(s)$.

From (1), by putting $a(0) = -1$, we get

$$R_{\varepsilon\varepsilon}(l) = \sum_{m=0}^{M} \sum_{n=0}^{M} a(m) \, a(n) \, R_{xx} \, (l-m+n) . \qquad (5)$$

By applying the relation (4) of 3.1.1 that is given by

$$R_{xx}(l) = \int_{-\frac{1}{2}}^{\frac{1}{2}} \exp{(i \, 2\pi \, g \, l)} \, q_{xx}(g) \, dg \qquad (6)$$

to (5) above, we get

$$R_{\varepsilon\varepsilon}(l) = \int_{-\frac{1}{2}}^{\frac{1}{2}} \exp{(i\,2\pi gl)}\, q_{\varepsilon\varepsilon}(g)\, dg$$

$$= \int_{-\frac{1}{2}}^{\frac{1}{2}} \exp{(i\,2\pi gl)} \left| \sum_{m=0}^{M} a(m) \exp{(-i\,2\pi gm)} \right|^2 q_{xx}(g)\, dg\,. \qquad (7)$$

Since $q_{\varepsilon\varepsilon}(g) = \sigma^2$, we get

$$q_{xx}(g) = \frac{\sigma^2}{\left| 1 - \displaystyle\sum_{m=1}^{M} a(m) \exp{(-i\,2\pi gm)} \right|^2}\,. \qquad (8)$$

Thus if the set of coefficients $a(m)$ is such that the corresponding prediction error $\varepsilon(s)$ is uncorrelated with the past values of $x(s)$, i.e., if the $a(m)$'s define the best linear predictor, and the variance $R_{\varepsilon\varepsilon}(0) = \delta^2$ is also given, the power spectral density function $q_{xx}(g)\,(-1/2 \leq g \leq 1/2)$ is defined uniquely. This is the reverse view of Wiener's discussion of the prediction problem and shows that the spectral density function can be determined by solving the prediction problem.

Suppose that the original phenomenon has continous time parameter and $x(s)$ are the values sampled with sampling interval Δt. In this case, if we restore Δt, from (7) of 3.1.1 we obtain

$$p_{xx}(f\,;\Delta t) = \Delta t\, q_{xx}(f\Delta t) \quad \left(-\frac{1}{2\Delta t} \leq f \leq \frac{1}{2\Delta t} \right). \qquad (9)$$

For the comparison of estimates of $q_{xx}(g)$ with different Δt as estimates of $p_{xx}(f)$ the estimates must first be transformed into the form $p_{xx}(f;\Delta t)$ given by (9).

To obtain the estimates of $a(m)$ based on a data set of finite length $\{x(s); s=1, 2, \ldots, N\}$ we replace $R_{xx}(l)$ in (2) by their estimates $C_{xx}(l)$ obtained from the data given in 3.1.4 and minimize the corresponding $\bar{\varepsilon}^2$. The estimates thus obtained will be denoted by $a_M(m)(m=1, 2, \ldots, M)$. They are obtained as the solution of the M-th order simultaneous linear equation

$$\sum_{m=1}^{M} C_{xx}(l-m)\, a_M(m) = C_{xx}(l) \quad (l=1, 2, \cdots\cdots, M)\,. \qquad (10)$$

In a real application we do not know the value of M in advance so we calculate $a_M(m)$'s for $M=1, 2, ..., L$, where L is a properly chosen upper bound for the order, and check the result. To do this, we have to solve equations of the form (10) repeatedly. However, by the algorithm to be explained below, the calculation for the case $M=L$ gives all the solutions for M smaller than L. For the selection of M we propose the minimization of FPE(M), an estimate of the final prediction error (FPE) that is defined as the expected mean squared prediction error when $a_M(m)$ is used for prediction.

Summarizing the above discussion we obtain the following procedure for estimating the *autoregressive model* defined by (1) with white noise $\varepsilon(s)$.

Autoregressive Model Fitting (Univariate case) (See Program 5.3.1.)
Assume that the data set $\{x(s\,\Delta t); s=1, 2, ..., N\}$ is given.
1) Subtract the mean value. i.e., calculate

$$\bar{x}=\frac{1}{N}\sum_{s=1}^{N} x(s\Delta t)$$

and define $x(s)$ by

$$x(s)=x(s\Delta t)-\bar{x} \quad (s=1, 2, \cdots\cdots, N).$$

2) For $l=0, 1, ..., L$, compute

$$C_{xx}(l)=\frac{1}{N}\sum_{s=1}^{N-l} x(s+l)\,x(s).$$

3) Put $a_0(m)=0$ ($m=1, 2, ..., L$), $\qquad \sigma^2(0)=C_{xx}(0)$.
4) For $M=0, 1, ..., L$, compute $a_M(m)(m=1, 2, ..., M)$ and $\sigma^2(M)$ by the following recursive procedure:

$$a_{M+1}(M+1)=(\sigma^2(M))^{-1}\left(C_{xx}(M+1)-\sum_{m=1}^{M} a_M(m)\,C_{xx}(M+1-m)\right)$$

$$a_{M+1}(m)=a_M(m)-a_{M+1}(M+1)\,a_M(M+1-m) \quad (m=1, 2, \cdots\cdots, M)$$

$$\sigma^2(M+1)=\sigma^2(M)\,(1-(a_{M+1}(M+1))^2).$$

5) At the same time, also compute

$$\text{FPE }(M)=(N+(M+1))(N-(M+1))^{-1}\sigma^2(M)$$

and adopt $a_M(m)$ corresponding to the M that gives the minimum of FPE(M) ($M=0, 1, ..., L$).

6) The estimate of the power spectral density $q_{xx}(g)$ ($-1/2 \leq g \leq 1/2$) is given by

$$\hat{q}_{xx}(g) = \frac{\sigma^2(M)}{\left| 1 - \sum_{m=1}^{M} a_M(m) \exp(-i 2\pi gm) \right|^2}.$$

By restoring the sampling interval Δt we get the estimate of $p_{xx}(f; \Delta t)$ by

$$\hat{p}_{xx}(f; \Delta t) = \hat{q}_{xx}(f \Delta t) \Delta t \quad \left(-\frac{1}{2\Delta t} \leq f \leq \frac{1}{2\Delta t} \right).$$

(See Program 5.4.1.)

Remarks:

1. As a rule of thumb for L we suggest $L=2N^{1/2}$ or $L=3N^{1/2}$. L should be kept strictly below $0.5N$. Practically, continue computation until FPE(M) shows a steady increase. In the case where FPE(M) continues to decrease identifinitely with increasing M, the power spectrum often shows sharp line like peaks or some dips.

2. The numerical accuracy of computation drops as M increases. If $\sigma^2(M)$ takes a negative value it means that the accuracy of the computation is completely lost. When the calculation is done with ten digits such a phenomenon often occurs when FPE(M) continues decreasing until M reaches 70–80.

3. When M is chosen by the above procedure, the evaluation of the accuracy of $a_M(m)$ is not yet precisely performed. (It is expected that an approximation of the variance matrix will be given by $N^{-1}\sigma^2(M)R_M^{-1}$, where R_M^{-1} denotes the inverse of the $M \times M$ matrix R_M whose (l, m) element is $C_{xx}(l-m)$.)

4. It is instructive to compare the estimate of the power spectrum obtained by the present procedure with that obtained by the method of 3.1.4. Once L is properly chosen, the present method produces the desired result almost automatically, without the difficulty of the determination of the lag length h in the previous method.

3.2.2 Multivariate autoregressive model

Consider the k-dimensional stationary time series $X(s)=(x_1(s), x_2(s), ..., x_k(s))'$, where $'$ denotes transpose. It is assumed that $x_i(s)$ represents the deviation from its mean value. The autoregressive representation is given by

$$X(s) = \sum_{m=1}^{M} A(m) \ X(s-m) + U(s) . \qquad (1)$$

Here $A(m)$ is a $k \times k$ matrix and $U(s) = (\varepsilon_1(s), \ \varepsilon_2(s), \ \ldots\ldots, \ \varepsilon_k(s))'$ is a k-dimensional white noise with zero mean vector and with variance covariance matrix given by

$$R_{\varepsilon_i \varepsilon_j}(l) = \delta_{l,0} \ \sigma_{ij} , \qquad (2)$$

where $\delta_{l,0} = 1$ for $l=0$, and $=0$ otherwise. $U(s)$ is called the innovation of $X(s)$. The covariance of $\varepsilon_i(s)$ and $x_j(s-m)(m=1, 2, \ldots)$ is 0 for any arbitary i and j. Let $A_{ij}(m)$ denote the (i, j)-th element of $A(m)$. Then (1) can be represented in the form

$$x_i(s) = \sum_{m=1}^{M} \sum_{j=1}^{k} A_{ij}(m) \ x_j(s-m) + \varepsilon_i(s) \quad (i=1, 2, \ldots\ldots, k) . \qquad (3)$$

If we put $A_{ij}(0) = -1$ for $i=j$, and $=0$ otherwise, then we get from (3)

$$R_{\varepsilon_i \varepsilon_j}(l) = \sum_{m=0}^{M} \sum_{n=0}^{M} \sum_{u=1}^{k} \sum_{v=1}^{k} A_{iu}(m) \ A_{jv}(n) \ R_{uv}(l-m+n) . \qquad (4)$$

Here, $R_{uv}(l)$ denotes the cross-covariance of $x_u(s+l)$ and $x_v(s)$. Using the relation between the cross-covariance function and the cross-spectral density given by

$$R_{yx}(l) = \int_{-\frac{1}{2}}^{\frac{1}{2}} \exp \ (i \ 2\pi f l) \ p_{yx}(f) \ df \qquad (5)$$

(4) can be rewritten as

$$\sigma_{ij} = \sum_{u=1}^{k} \sum_{v=1}^{k} A_{iu}(f) \ p_{uv}(f) \ \overline{A_{jv}(f)} \quad (i, j=1, 2, \ldots\ldots, k) , \qquad (6)$$

where

$$A_{ih}(f) = \sum_{m=0}^{M} A_{ih}(m) \ \exp \ (-i \ 2\pi f m) . \qquad (7)$$

Let Σ, $A(f)$ and $P(f)$ denote matrices whose (i, j)-th elements are given by σ_{ij},

$A_{ij}(f)$ and $P_{ij}(f)$ (cross-spectral density of x_i and x_j), respectively. Then (6) can be rewritten as

$$\Sigma = A(f)\, P(f)\, \overline{(A(f))}' \tag{8}$$

and thus, provided that $(A(f))^{-1}$ exists, we get

$$P(f) = (A(f))^{-1}\, \Sigma\, ((\overline{A(f)})')^{-1}. \tag{9}$$

Here $A(f)$ can be represented as

$$A(f) = \sum_{m=0}^{M} A(m)\, \exp\,(-i\, 2\pi\, fm), \tag{10}$$

where $A(0) = -I$ and I is a $k \times k$ identity matrix.

Result (9) shows that once an autoregressive representation (11) is given, the spectrum of the k-dimensional stationary time series is completely specified. All the quantities needed for the statistical spectrum analysis of a time invariant linear system are given by $P(f)$. Thus the result (9) shows that if the prediction problem of a time series is solved, then the problem of determining the characteristics of the corresponding linear system is also solved completely.

For a given M, denote by $A_{Mij}(m)$ the value of $A_{ij}(m)$ that minimizes the mean square of $\varepsilon_i(s)$ that is obtained from $x_j(s)$'s through relation (3), and define the matrix of these values by $A_M(m)$. If the auto and cross-covariances $R_{ij}(l)(i, j = 1, 2, ..., k; l = 1, 2, ..., M)$ of x_i and $x_j(i, j = 1, 2, ..., k)$ are given, $A_M(m)$ can be obtained by solving the following simultaneous linear equations for $i = 1, 2, ..., k$

$$\sum_{m=1}^{M} \sum_{j=1}^{k} A_{Mij}(m)\, R_{jh}(l-m) = R_{ih}(l) \quad (h=1, 2, \cdots, k; l=1, 2, \cdots, M). \tag{11}$$

Since (11) is a $(M \times k)$-dimensional simultaneous first-order equation, it is very difficult to solve numerically when k and M are large. Especially when we have to examine various values of M, as is always the case in practical applications, the necessary amount of calculation becomes extremely large. Fortunately, as will be explained below, for this case also a computationally very efficient procedure is available, just as in the univariate case. By using this procedure, all the solutions for $M = 1, 2, ..., L$ can be obtained during the process of solving the one equation for $M = L$.

Just as in the univariate case, we can determine the multivariate autoregressive model on the basis of observed data by using $C_{ij}(l)$ in place of $R_{ij}(l)$ and determining M by using a quantity corresponding to FPE(M). This provides the basis for the design procedure of a control system to be discussed later in this chapter.

Autoregressive Model Fitting (Multivariate case) [9] (See Program 5.3.2.)

Assume that the data set $\{x_1(s\Delta t), x_2(s\Delta t), ..., x_k(s\Delta t); s=1, 2, ..., N\}$ is given.

1) For $i=1,2, ..., k$, calculate

$$\bar{x}_i = \frac{1}{N} \sum_{s=1}^{N} x_i(s\Delta t)$$

and re-define $x_i(s)$ by

$$x_i(s) = x_i(s\Delta t) - \bar{x}_i \quad (s=1, 2, \cdots\cdots, N).$$

2) For $i, j = 1, 2, ..., k$, calculate

$$C_{ij}(l) = \frac{1}{N} \sum_{s=1}^{N-l} x_i(s+l) \, x_j(s) \quad (l=0, 1, \cdots\cdots, L).$$

Here, $C_{ij}(0) = C_{ji}(0)$. We define a $k \times k$ matrix $C(l)$ with $C_{ij}(l)$ as the (i, j)-th element. The calculations below are all concerned with $k \times k$ matrices.

3) Let $A_0(m) = O$ (zero matrix)$(m=1, 2, ..., L)$.

4) Calculate $A_M(m)(m=1, 2, ..., M)$ and d_M for $M=0, 1, ..., L$, by the following recursive relations (see remark 3 below):

$$d_M = C(0) - \sum_{m=1}^{M} A_M(m) \, C'(m)$$

$$e_M = C(M+1) - \sum_{m=1}^{M} A_M(m) \, C(M+1-m)$$

$$f_M = C(0) - \sum_{m=1}^{M} B_M(m) \, C(m)$$

$$D_M = e_M f_M^{-1}$$

$$E_M = e_M' \, d_M^{-1}$$

$$A_{M+1}(m) = A_M(m) - D_M B_M (M+1-m) \quad (m=1, 2, \cdots\cdots, M)$$

$$\qquad\qquad = D_M \qquad\qquad\qquad\qquad\qquad (m=M+1)$$

$$B_{M+1}(m) = B_M(m) - E_M A_M (M+1-m) \quad (m=1, 2, \cdots\cdots, M)$$
$$\qquad\qquad = E_M \qquad\qquad\qquad\qquad\qquad\qquad (m=M+1)$$

where, ' denotes transpose.

5) Also compute

$$\text{MFPE } (M) = \left(1 + \frac{Mk+1}{N}\right)^k \left(1 - \frac{Mk+1}{N}\right)^{-k} \| d_M \|$$

and adopt the value of M that gives the minimum of MFPE(M). Here, $\|d_M\|$ denotes the determinant of d_M. d_M gives an estimate of Σ, the covariance matrix of the white noise.

6) $\hat{P}(f)$, an estimate of the power and cross spectral density matrix $P(f)$, can be obtained by (9) if we use the above obtained M and replace $A(m)$ and Σ of (9) and (10) by $A_M(m)$ and d_M, respectively. The necessary re-interpretation for the restoration of Δt is the same as in the univariate case. (See Program 5.4.2.)

Remarks:

1. Keep L below $N/(5k)$. As a rule of thumb $2N^{1/2}/k$ or $3N^{1/2}/k$ may be used as L. However, L must be strictly below $N/(2k)$ to insure the appropriate behavior of MFPE(M).

2. By using $\hat{P}(f)$ obtained by (6) for the computation discussed in 3.1.4, we can obtain the estimates of the frequency response functions and coherencies. However, we can also use the relation

$$(P(f))^{-1} = \overline{(A(f))}' \, \Sigma^{-1} \, (A(f)) \tag{12}$$

which is obtained from (9). Let $G_{ij}(f)$ denote the frequency response function of the system with $x_j(s)$ as input and $x_i(s)$ as output, then

$$G_{ij}(f) = -\frac{(P(f))_{ij}^{-1}}{(P(f))_{ii}^{-1}} \tag{13}$$

where $(P(f))_{ij}^{-1}$ denotes the (i, j)-th element of $P(f)^{-1}$. This relation represents the content of the matrix calculation for the estimation of the frequency response functions and coherencies discussed in 3.1.4. (Put $i=0$ to get result of 3.1.4.)

3. For the computation of d_M^{-1} or f_M^{-1} that appear in stage 4) or for the computation of $\|d_M\|$ of stage 5) the use of a subroutine based on the computational procedure to be discussed in the next subsection 3.2.3 is advisable.

3.2.3 Simulation procedure

In this section, we discuss some technical points of computation required for simulation experiments. Since this section is not directly related to the main stream of this book, readers who are less concerned with algorithmic aspects may skip this subsection.

The autoregressive model is quite suited for the purpose of simulation of a stationary multivariate time series. To obtain a realization of $X(s)(s=0, 1, 2, ...)$ from the relation

$$X(s) = \sum_{m=1}^{M} A(m) \, X(s-m) + U(s),\tag{1}$$

which was given in (1) of 3.2.2, a set of initial values $X(-1), X(-2), ..., X(-M)$ and a realization of the white noise sequence $U(0), U(1), ..., U(s)$ is all that is needed. Here, of course, we assume that the $k \times k$ matrices $A(m)(m=1, 2, ..., M)$ are given. $U(s)$ is a k-dimensional white noise with mean 0 (zero vector) and covariance matrix Σ. When we are concerned with the behavior of $X(s)$ at sufficiently large s, where the influence of the initial conditions can be ignored, the initial values of $X(s)$ may all be set equal to 0. The computational problem is then to obtain the realization of $U(s)$.

For this purpose we use the following decomposition, often called the Cholesky decomposition,

$$\Sigma = LL'\tag{2}$$

where L is a lower triangular matrix

$$L = \begin{bmatrix} l_{11} & 0 & \cdots\cdots & 0 & 0 \\ l_{21} & l_{22} & \cdots\cdots & 0 & 0 \\ \vdots & \vdots & & \vdots & \vdots \\ l_{k-1,1} & l_{k-1,2} & \cdots\cdots & l_{k-1,k-1} & 0 \\ l_{k1} & l_{k2} & \cdots\cdots & l_{k,k-1} & l_{kk} \end{bmatrix}\tag{3}$$

with (i, j)-th element l_{ij} such that $l_{ij}=0$ for $i<j$. When such a matrix L is given, the realization of $U(s)$ can be obtained by using the following equation

$$U(s) = LO(s)\tag{4}$$

where $O(s)=(o_1(s), o_2(s), ..., o_k(s))'$ denotes a vector of k random numbers which are independent realizations of a random variable with mean 0 and variance 1.

To get L we first calculate M which satisfies

$$\Sigma^{-1} = M' M \tag{5}$$

and then get $L = M^{-1}$. To obtain M we first construct a matrix composed of Σ and a $k \times k$ unit matrix and given by

$$\begin{bmatrix} \sigma_{11} & \sigma_{12} & \sigma_{13} & \cdots\cdots & \sigma_{1k} & 1 & 0 & 0 & \cdots\cdots & 0 \\ \sigma_{21} & \sigma_{22} & \sigma_{23} & \cdots\cdots & \sigma_{2k} & 0 & 1 & 0 & \cdots\cdots & 0 \\ \sigma_{31} & \sigma_{32} & \sigma_{33} & \cdots\cdots & \sigma_{3k} & 0 & 0 & 1 & \cdots\cdots & 0 \\ \vdots & \vdots & \vdots & & \vdots & \vdots & \vdots & \vdots & & \vdots \\ \sigma_{k1} & \sigma_{k2} & \sigma_{k3} & \cdots\cdots & \sigma_{kk} & 0 & 0 & 0 & \cdots\cdots & 1 \end{bmatrix}. \tag{6}$$

First, divide the elements of the first row by $\sigma_{11}^{1/2}$ and then divide the elements of the first column by the same quantity. Next, multiply the 1st row by the value of the $(i, 1)$-th element and subtract this from the i-th row $(i=2, 3, ..., k)$. In this way the elements of the 1st column are transformed in such a away that the $(1, 1)$-th element becomes 1 and the rest 0.

To the resulting $(k-1) \times (2k-1)$ matrix composed of the rows and columns obtained by eliminating the first row and column, apply the same type of transformation. This is equivalent to the following procedure. Redefine σ_{22} by the $(2,2)$-th element of the $k \times 2k$ matrix resulting from the transformation of the first row and column explained above. Divide the second row and then the second column by $\sigma_{22}^{1/2}$. Multiply the second row by the value of the $(i,2)$-th element and subtract it from the i-th row $(i=3, 4, ..., k)$.

Proceed to the k-th row repeating similar operations. When the (k,k)-th element is transformed into 1 the original $k \times k$ identity matrix on the right hand side of matrix (6) is transformed into the desired lower triangular matrix M.

In an actual computation we do not need the $k \times k$ unit matrix and operate only on the matrix Σ on the left hand side of matrix (6). The computation then proceeds as follows:

1) Replace the $(1,1)$-th element σ_{11} by $\sigma_{11}^{-1/2}$.

2) Multiply the elements of the 1st row, except that of the 1st column, by this new $(1,1)$-th element.

3) Change the sign of the $(1,1)$-th element and multiply the elements of the 1st column, except that of the 1st row, by the new $(1,1)$-th element.

4) For $i=2,3, \ldots, k$, repeat the following operations. Multiply the elements of the 1st row, except that of the 1st column, by the $(i, 1)$-th element and add them to the corresponding elements of the i-th row $(i=2, 3, \ldots, k)$, respectively. Then multiply the $(i,1)$-th element by the $(1,1)$-th element and let this be the new $(i,1)$-th element

5) Denote the $(2,2)$-th element of the matrix thus obtained by σ_{22} and replace σ_{22} by $\sigma_{22}^{-1/2}$.

6) Multiply the elements of the 2nd row, except for the $(2,2)$-th element, by the value of the $(2,2)$-th element.

7) Change the sign of the $(2,2)$-th element and multiply the elements of the second column, on and below the 3rd row, by this new $(2,2)$-th element.

8) Multiply the elements of the 2nd row, except that of the 2nd column, by the $(i,2)$-th element and add them to the corresponding elements of the i-th row for $i=3, 4, \ldots, k$. Multiply the $(i,2)$-th element by the $(2,2)$-th element and let this be the new $(i,2)$-th element

9) Proceed similarly until σ_{kk} is replaced by $\sigma_{kk}^{-1/2}$. At this point the triangular matrix on the lower half of the $k \times k$ matrix gives the desired matrix M.

10) In the matrix obtained through steps (1)–(9), replace the diagonal elements by their reciprocals, respectively. Then the triangular matrix on and above the diagonal of this $k \times k$ matrix gives $(M')^{-1}=L'$.

If necessary Σ^{-1} can be obtained easily by using (5) and M obtained by the above calculation. If we first define $\mathrm{DET}=1.0$ and in steps 1), 5), 9), etc, replace this by $\mathrm{DET} \times \sigma_{ii}$ before replacing σ_{ii} by $\sigma_{ii}^{-1/2}$ then we get $\mathrm{DET}=\|\Sigma\|$, the value of the determinant of Σ, when step 9) is completed. By taking the transpose of L' obtained at step 10) we get L.

When the Gaussian property (normality) is required in the simulation, simply simulate the original input $z_1(s), z_2(s), \ldots, z_k(s)$ from random variables following the Gaussian distribution with mean 0 and variance 1 and put

$$\varepsilon_i(s) = \sum_{j=1}^{i} L_{ij} \, z_j(s) \tag{7}$$

where L_{ij} denotes the (i, j)-th element of L. The vector $(\varepsilon_1(s), \varepsilon_2(s), \ldots, \varepsilon_k(s))'$ gives the realization of $U(s)$. By the autoregressive model the value of $X(s)$ is obtained as the linearly weighted sum of the $U(s-m)'s$ $(m=0, 1, \ldots)$. When significant weights are distributed over many values the distribution of $X(s)$ becomes close to Gaussian even if a random number uniformly distributed over $[-3^{1/2}, 3^{1/2}]$ is used as $z_i(s)$.

In any case, the realization of $X(s)$ $(s=1, 2, \ldots, N)$ can be obtained by substituting the realized value of $U(s)$ $(s=1, 2, \ldots, N)$ into (1), if only the set of values of $X(s)$ at the first M time points is specified; for example, $X(0)=X(-1)=\ldots=X(-M+1)=0$. When the covariance matrices of $X(s)$ are given, by using a proper order M and fitting an autoregressive model of the form (1) we can simulate the behavior of the time series with covariance matrices identical to the given matrices at least up to the M-th lag. For the fitting of the autoregressive model the method of 3.2.2 is applicable. However, in this case, if the covariance matrices are given theoretically, the determination of M can be done only by a subjective judgement of the accuracy of approximation given by the model.

The above method of realizing the observed values of $U(s)$ is also used in the simulation of the optimal control system. For the programs on this subject see Program 5.5.2 OPTSIM and Program 5.5.3 WNOISE given in Chapter 5.

3.3   Analysis of Feedback Systems

In 3.2 it was mentioned that, in the discussion of a stationary time series, the choice between the spectral density and the optimal linear predictor as the starting point depended mainly on the practical convenience. However, when we actually try to obtain results from observed data, a great difference between the two becomes apparent. It can be seen that not only the numerical accuracies of estimates but also even the feasibility of estimation depends on the approach. This point is quite clear when we analyze the feedback system which will be discussed in this section. In the case of a feedback system explained below, it is practically impossible to analyze the properties of a system only by using the frequency domain approach based on spectral densities. For the analysis of the physical realizability of the system, the property that the influence of the input appears in the output with a time lag, plays a decisive role and thus the time domain approach, which can handle this property directly, becomes by far the more advantageous.

First we will give a brief explanation of the kind of feedback system to be considered here. The simplest case is shown in Fig. 3.3-1. This system is composed of two subsystems A and B that are driven by the inputs $x(s)$ and $y(s)$, respectively. It is assumed that A and B are both linear and time invariant. The variables $y(s)$ and $x(s)$ are given as the sums of the output of A and noise $u(s)$ and the output of B and noise $v(s)$, respectively. The whole system is driven by the noise sources $u(s)$ and $v(s)$ and it is assumed that if both of them vanish the system comes to a standstill. Further, it is assumed that $\{u(s)\}$ and $\{v(s)\}$ are statistically independent. This is equivalent to assuming that each subsystem has its own noise source. Although this is an example of the utmost simplicity, it seems that many real systems have this type of structure. The following examples are of this form.

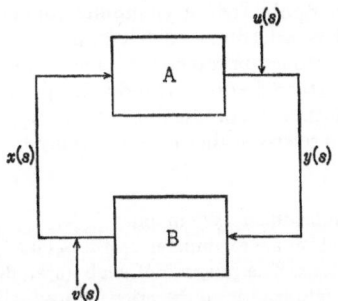

Fig. 3.3-1   A feedback system with noise source.

1) Driving a car

Consider a car being driven on a given straight course drawn on uneven ground. Let $x(s)$ denote the angle of rotation of the steering wheel of the car and $y(s)$ the deviation of the car from the course. In this case, A represents the characteristic of the car and B the characteristic of the driver. The effect of the roughness of the road surface is represented by $u(s)$ and that of the variability of the driver's response is represented by $v(s)$. If the amplitudes of the fluctuations of $x(s)$ and $y(s)$ are both kept sufficiently small, A and B may be approximated by linear systems.

Here the behavior of the driver is oriented towards the reduction of the value of $y(s)$, the amount of deviation from the course, by feeding back the value of $y(s)$ to $x(s)$.

A similar model could be considered for the description of the action of pilot trying to retain the horizontal stability of an air plane. Many systems controlled by a human operator could be described by this model. When $y(s)$ denotes the temperature at some point in a production process and $x(s)$ the rate of fuel consumption and an operator is manipulating $x(s)$ to minimize the fluctuation of $y(s)$ the system can similarly be expressed. A real example of this kind of system is given in 4.3.

2) Fluctuation of the hog price

Let $x(s)$ denote the amount of hog production and $y(s)$ the price, respectively, measured from the mean values. Assume that A represents the action of the market and B the reaction of the producer. The noises $u(s)$ and $v(s)$ can then be regarded as unexpected movements of the price and output, respectively. This is an extremely simplified model which may be difficult to accept from the standpoint of an economic theory. Nevertheless, in agricultural markets it is easily observed that the price of a product goes down when the amount of production is increased, the production decreases with the fall of the price, and then the drop in production produces a rise in price to eventually stimulate increase in production. This shows that many economic systems have the characteristic that can essentially be represented by the present simple model.

3.3.1 The purpose of feedback system analysis

The above market model example is useful in clarifying the purpose of feedback system analysis. The purpose of such an analysis may usually be considered to be the stabilization of the price or production.

To attain this purpose, it is most important first to recognize the weakest point of the system. For this reason, it is necessary to understand the characteristics of $u(s)$ and $v(s)$, the noise sources, as well as the characteristics of A and B. The goal of feedback system analysis cannot be achieved unless we know the statistical characteristics of the noise sources $u(s)$ and $v(s)$, along with the characteristics of the system A and B. Once this information is obtained, it becomes clear to which point attention should be paid to handle the situation. For example, the characteristic of the noise $u(s)$ may provide a clue to its generation mechanism and this may suggest a procedure that leads to a significant reduction of its power. When the original $u(s)$ has significant power, it often happens that no significant improvement can be produced by simply manipulating the characteristics of the system such as A and B. This point was already mentioned in Section 1.4 where we discussed the purpose of the analysis of a system.

To develop an understanding of a process showing irregular fluctuations such as $x(s)$ and $y(s)$, it is essential to recognize the effect of the structure of the system and that of the noise source. Where necessary we must proceed further in the analysis of the present noise source by regarding it as a system driven by another noise source, until we recognize the part of the process that can most effectively be modified. This analysis does not necessarily lead to a decomposition of the form of Fig. 3.3-1 nor is it always possible to express it mathematically. Nevertheless, extremely effective results may sometimes be obtained by conceptually performing this type of analysis based on previous experience and knowledge of the field of application.

We are about to start the discussion of the statistical method for the analysis of a system described by Fig. 3.3-1 and its generalizations. In the present subsection we reconfirmed the fact that the purpose of the analysis of a feedback system is to clarify the characteristics of the system and noise sources. This was to ensure that we would not forget the real road to success by putting too much faith in the formal use of a statistical procedure.

3.3.2   Spectral analysis of feedback systems [10]

First of all we must understand that, in the analysis of a feedback system such as the one described by Fig. 3.3-1, the method of spectral analysis discussed in 3.1 is not useful for the handling of real data, although it may be useful for the handling of theoretical problems. For example, suppose that we calculated the frequency response function of A in Fig. 3.3-1 from $p_{yx}(f)/p_{xx}(f)$ by applying the method of 3.1.2 based on the cross spectral density functions. $z(s)$ and $n(s)$ in (6) of 3.1.2 correspond to $y(s)$ and $u(s)$ of Fig. 3.3-1, respectively. Formula (13) of 3.1.2 was based on the basic relation (7) or (8) of 3.1.2, and the assumption that the input $x(s)$ and noise $n(s)$ are uncorrelated. In the present case, the noise $u(s)$ produces an effect on $x(s)$ through B and thus $x(s)$ and $u(s)$ are generally not uncorrelated. Thus $p_{yx}(f)/p_{xx}(f)$ does not always give the frequency response function of A. Generally, the quantity represents a complicated weighted average of the frequency response functions of A and B. When $v(s)$ has positive power spectral density while $u(s)$ is vanishing, $p_{yx}(f)/p_{xx}(f)$ is identical to the frequency response function of A. But this is an exceptional case. The Fourier transform of $p_{yx}(f)/p_{xx}(f)$ will generally extend over both positive and negative sides of the time axis, even when the original A and B are physically realizable, i.e., with impulse response functions extending only on the positive portion of the time axis. At present, for the case of Fig. 3.3-1, there is no effective method to determine the frequency response function of A or B by the frequency domain approach. Taking into account the fact that, as the preceding discussion has shown, many practically important problems could be represented by the scheme of Fig. 3.3-1, this must be considered as a serious defect of the spectral method of analysis [11].

As was just explained, for the analysis of a feedback system of the form of Fig. 3.3-1, no useful result can directly be obtained by the cross-spectral method explained in and after 3.1.2. However, if we apply the autoregressive representation of the stationary time series discussed in 3.2 to this problem, we can easily get a practical solution. Therefore the spectral analysis of a feedback system can be realized in the time domain and then the result can be transformed into the frequency domain representation which is more amenable to physical interpretation.

We now discuss a method for the measurement of the characteristics of A or B, where a crucial assumption is that the noise sourses $u(s)$ and $v(s)$ are uncorrelated. This assumption will be discussed separately later. When M is chosen to be sufficiently large, the system in Fig. 3.3-1 can be represented, at least approximately, by

$$y(s) = \sum_{m=1}^{M} a(m) \; x(s-m) + u(s)$$

$$x(s) = \sum_{m=1}^{M} b(m) \; y(s-m) + v(s).$$

(1)

For a system with continuous time parameter t it is assumed that the length of the interval between consecutive observations, Δt, is chosen to be sufficiently small. Here, $\{a(m)\}$ and $\{b(m)\}$ denote the impulse response functions of A and B, respectively. There are situations where $a(0)$ or $b(0)$ must be considered, but, for the sake of simplicity, we now concentrate on the model given by (1).

When a record of observations of $x(s)$ and $y(s)$ is given by $\{x(s), y(s): s = -M+1, -M+2, \ldots, N\}$, what would be the procedure for obtaining the estimate of $\{a(m)\}$? When $x(s)$ and $y(s)$ are measured as deviations from their means, $u(s)$ and $v(s)$ may be considered as random variables with zero means. In this case one may consider the application of the method of least squares. By this method we obtain $a_0(m)$, the values of $\alpha(m)$ that minimize the sum of squares of the residuals

$$\sum_{s=1}^{M} \left\{ y(s) - \sum_{m=1}^{M} \alpha(m) \; x(s-m) \right\}^2$$

(2)

and adopt $\{a_0(m)\}$ as the estimate of the inpulse response function $\{a(m)\}$. However, this method gives a biased result, no matter how large the number of data points N may be, except for the special case where the correlations between $u(s)$ and $x(s-1)$, $x(s-2)$, ..., $x(s-M)$ are vanishing.

When there is a feedback from $y(s)$ to $x(s)$ through B the above condition cannot be expected to hold, unless $u(s)$ is a white noise. This fact must be remembered when the method of least squares is applied to the estimation of the dynamics of a system. When there is no feedback, the least squares estimate $a_0(m)$ will converge to $a(m)$ "with probability 1" (we will omit this expression hereafter) as N goes to infinity, provided that $\{x(s), y(s)\}$ is ergodic. However for a finite N this $a_0(m)$ cannot always be considered to be a good estimate in the sense of the mean square of the deviation from $a(m)$. The least squares method is certain to produce a good result only when $u(s)$ of (1) forms a sequence of independent identically distributed random variables.

When $u(s)$ is not a white noise, as will be discussed below, an efficient procedure of estimation can be obtained by transforming $u(s)$ into a white noise. The method is applicable to the case where feedback from $y(s)$ to $x(s)$ exists. From the discussion of 3.2.1, a practically useful autoregressive representation of $u(s)$ and $v(s)$ can be given by

$$u(s) = \sum_{l=1}^{L} c(l) \, u(s-l) + \xi(s)$$

$$v(s) = \sum_{l=1}^{L} d(l) \, v(s-l) + \eta(s) \tag{3}$$

where L is a properly chosen sufficiently large integer. Here $\xi(s)$ and $\eta(s)$ denote white noises which are mutually uncorrelated. As $u(s)$ and $v(s)$ are stationary, the system (3) that generates $u(s)$ and $v(s)$ from $\xi(s)$ and $\eta(s)$ must be stable and $u(s)$ and $v(s)$ are given as limits of linear combinations of the present and past values of $\xi(s)$ and $\eta(s)$, respectively. From this, we see that the past values of $x(s)$ and $y(s)$ such as $x(s-1), y(s-2),\ldots$ are uncorrelated with $\xi(s)$. This solves the difficulty of the application of the least squares method.

By using the first equation of (3), we can transform $u(s)$ into $\xi(s)$ which is a white noise. Applying this same transformation to both sides of the first equation of (1), we get

$$y(s) - \sum_{l=1}^{L} c(l) \, y(s-l) = \sum_{m=1}^{M} a(m) \left(x(s-m) - \sum_{l=1}^{L} c(l) \, x(s-m-l) \right)$$

$$+ u(s) - \sum_{l=1}^{L} c(l) \, u(s-l), \tag{4}$$

from which it follows that

$$y(s) = \sum_{l=1}^{L} c(l) \, y(s-l) + \sum_{m=1}^{M+L} A_m \, x(s-m) + \xi(s), \tag{5}$$

where $a(m)=0(m>M)$ and

$$A_1 = a(1)$$

$$A_m = a(m) - \sum_{l=1}^{m-1} c_l \, a(m-l) \quad (m=2, 3, \cdots\cdots, M+L) \tag{6}$$

where

$$c_l = c(l) \qquad (l=1, 2, \cdots\cdots, L),$$
$$= 0 \qquad \text{for} \quad l>L \, .$$

Since $\xi(s)$ is uncorrelated with $y(s-l)$, $x(s-m)\ldots$, when the least squares method is applied to (5), the resulting estimates converge to $c(l)$ and A_m respectively, when the number of observations N is increased indefinitely.

When N is finite but sufficiently large, the estimates $c_0(l)$ and A_{0m} will be close to $c(l)$ and A_m, respectively.

Rewriting (6) we get

$$a\ (1)=A_1$$

$$a\ (m)=A_m+\sum_{l=1}^{m-1} c_l\, a(m-l)\quad (m=2, 3, \cdots\cdots, M)\ . \tag{7}$$

Accordingly, by replacing A_m and $c_l(m=1, 2, ..., M; l=1, 2, ..., L)$ by their estimates A_{0m} and $c_0(l)$ we can obtain $a_1(m)$, the estimates of $a(m)$ for $m=1, 2, ..., M$, by

$$a_1(1)=A_{01}$$

$$a_1(m)=A_{0m}+\sum_{l=1}^{m-1} c_0(l)\, a_1(m-l)\quad (m=2, 3, \cdots\cdots, M), \tag{8}$$

where $c_0(l)=0(l>L)$. Since $c_0(l)$ and A_{0m} converge to $c(l)$ and A_m when N tends to infinity it is obvious from (7) that $a_1(m)$ will converge to $a(m)$. Thus we can see, that even under the existence of the feedback we can obtain an estimate of $\{a(m)\}$ that converges to the true impulse response function as N increases. For the case where there is no feedback the above whitening transformation of $u(s)$ provides an estimate which is more efficient than that obtained by the direct application of the least squares method. This problem is discussed in detail in [12].

If we use the estimate of $c(l)$ and that of the mean square of $\xi(s)$ (the mean square of the residuals when the least squares method is applied to (5)) in place of $a(m)$ and σ^2 of (8) of 3.2, with $M=L$, we get an estimate of the power spectral density function of $u(s)$. Thus, for the case of Fig. 3.3-1, we now have a procedure to determine the characteristics of the system and the noise source.

By extending the model of Fig. 3.3-1, further as is discussed in [10], we obtain the extension of (1) to the case of k-dimensional variable $(x_1(s), x_2(s), ..., x_k(s))$ given by

$$x_i(s)=\sum_{j=1}^{k}\sum_{m=1}^{M} a_{ij}(m)\ x_j(s-m)+u_i(s)\quad (i=1, 2, \cdots\cdots, k)\ . \tag{9}$$

Here, $\{a_{ij}(m)\}$ denotes the impulse response function of $x_i(s)$ to the input $x_j(s)$. We consider that $x_i(s)$ is composed of the effects of the variables $x_j(s)(j\neq i)$ and its own noise $u_i(s)$ and thus put $a_{ii}(m)=0$. By drawing a figure corresponding to Fig. 3.3-1 for (9), we can see that this gives a natural model for a feedback system.

Generally, the representation of the system is not unique. However, (9) gives a particular representation where u_i represents the fluctuation of x_i after all the influences of the fluctuations of $x_j(j \neq i)$ were suppressed. In this sense, u_i represents the fluctuation originating in x_i. Estimates of $a_{ij}(m)$ can be obtained by using the whitening transformation of $u_i(s)$, just as in the two-dimensional case.

If we apply the least squares method directly to representation (5), the matrix required for the computation will be of magnitude about $k(M+L)$ $\times k(M+L)$ and the calculation becomes extremely time and space consuming. Also, the determination of L and M poses a serious practical difficulty. However, these problems can be solved easily by applying the method of fitting the multivariate autoregressive model, given in 3.2.

Before discussing this method, we will consider the spectral analysis of the feedback system of (9). The frequency response function $a_{ij}(f)$ of $x_i(s)$ to the input $x_j(s)$ is given by

$$a_{ij}(f) = \sum_{m=1}^{M} a_{ij}(m) \exp(-i\, 2\pi f m). \tag{10}$$

The system given by (9) is a feedback system within which $x_j(s)$ is connected to $x_i(s)$ by an element having the frequency response function $a_{ij}(f)$ and each $x_i(s)$ has its own noise source $u_i(s)$. Thus, $x_i(s)$ can be expressed as a sum of the influences of $u_j(s)$'s. Under the assumption that $u_j(s)$'s are mutually uncorrelated, the power spectral density function $p_{ii}(f)$ of $x_i(s)$ can be expressed as a sum of the power contributions from each $u_j(s)$. If we denote the power spectral density function of $u_j(s)$, by $p(u_j)(f)$, then, as the influence of $x_j(s)$ on $x_i(s)$ is generated by the frequency response function $b_{ij}(f)$ of $x_i(s)$ to the input $x_j(s)$ within the closed loop system given by (9), we get

$$p_{ii}(f) = \sum_{j=1}^{k} |b_{ij}(f)|^2 \, p(u_j)(f), \tag{11}$$

where $b_{ij}(f)$ is given by

$$\begin{bmatrix} b_{11}(f) & b_{12}(f) & \cdots & b_{1k}(f) \\ b_{21}(f) & b_{22}(f) & \cdots & b_{2k}(f) \\ \vdots & \vdots & & \vdots \\ b_{k1}(f) & b_{k2}(f) & \cdots & b_{kk}(f) \end{bmatrix} = \begin{bmatrix} 1-a_{11}(f) & -a_{12}(f) & \cdots & -a_{1k}(f) \\ -a_{21}(f) & 1-a_{22}(f) & \cdots & -a_{2k}(f) \\ \vdots & \vdots & & \vdots \\ -a_{k1}(f) & -a_{k2}(f) & \cdots & 1-a_{kk}(f) \end{bmatrix}^{-1} \tag{12}$$

If we define $q_{ij}(f)$ by

$$q_{ij}(f) = |b_{ij}(f)|^2 p(u_j)(f) \tag{13}$$

this represents the contribution of $u_j(s)$ to the power spectral density of $x_i(s)$ at the frequency f. Accordingly the relative power contribution is given by

$$r_{ij}(f) = \frac{q_{ij}(f)}{p_{ii}(f)} \tag{14}$$

and the commulative relative power contribution is given by

$$R_{ij}(f) = \sum_{h=1}^{j} r_{ih}(f) \quad (j=1, 2, \cdots\cdots, k). \tag{15}$$

When these quantities are graphically represented the pattern of the contributions of the noise sources to the system behavior becomes quite clear. This provides useful information in the analysis of the feedback system explained in 3.3.1. A practical example of this type of analysis will be discussed in 4.3 in relation to the analysis of the cement kiln process.

Through the fitting of a multivariate autoregressive model, the necessary quantities for the spectral analysis can be obtained quite easily. As in (5) consider the application of the whitening operation to $u_i(s)$ of (9) to obtain

$$x_i(s) = \sum_{l=1}^{L} c_i(l) \, x_i(s-l) + \sum_{\substack{j=1 \\ j \neq i}}^{k} \sum_{m=1}^{M+L} A_{ijm} \, x_j(s-m) + \varepsilon_i(s) \quad (i=1, 2, \cdots\cdots, k), \tag{16}$$

where

$$u_i(s) = \sum_{l=1}^{L} c_i(l) \, u_i(s-l) + \varepsilon_i(s) \tag{17}$$

and $\varepsilon_i(s)$ denotes a white noise. Obviously (16) is a special case of the general form

$$x_i(s) = \sum_{m=1}^{M} \sum_{j=1}^{k} A_{ij}(m) \, x_j(s-m) + \varepsilon_i(s) \quad (i=1, 2, \cdots\cdots, k). \tag{18}$$

Note that M of (18) corresponds to $M+L$ of (16). The formula (18) is identical to (3) of 3.2.2 which defines the multivariate autoregressive model. The only difference is that in the present model it is assumed that the $\varepsilon_i(s)$'s are mutually uncorrelated. This assumption allows us to obtain the characteristics of the noise source and the system separately in the form of $c_i(l)$ and $a_{ij}(m)$ from the result of fitting of the general model (18).

Suppose that an expression of the form (18) has been obtained. Then from (9) of 3.2.2 the matrix $P(f)$ of the spectral density functions of $x_i(s)(i=1, 2, ..., K)$ is given by

$$P(f) = (A(f))^{-1} \Sigma \left(\overline{(A(f))'} \right)^{-1}, \tag{19}$$

where Σ denotes the covariance matrix of the $\varepsilon_i(s)$'s. The (i, j)-th element of Σ, denoted by σ_{ij}, is the covariance of $\varepsilon_i(s)$ and $\varepsilon_j(s)$. Particularly σ_{ii} is the variance of $\varepsilon_i(s)$. The (i, j)-th element $p_{ij}(f)$ of $P(f)$ gives the cross-spectral density function of $x_i(s)$ and $x_j(s)$. When $j=i$ it gives the power spectral density function of $x_i(s)$. Also, from (19) of 3.2.2, $A(f)$ is given by

$$A(f) = -\left(I - \sum_{m=1}^{M} A(m) \exp(-i2\pi fm) \right), \tag{20}$$

where I is a $k \times k$ identity matrix, and $A(m)$ is a $k \times k$ matrix with $A_{ij}(m)$ as its (i, j)-th element. The negative sign in front of the parentheses of the right hand side of (20) may be ignored.

Under the assumption that the $\varepsilon_i(s)$'s are mutually uncorrelated we have $\sigma_{ij}=0(i \neq j)$ and from (19) we get

$$p_{ii}(f) = \sum_{j=1}^{k} |(A(f))^{-1}_{ij}|^2 \sigma_j^2 \tag{21}$$

where $(A(f))^{-1}_{ij}$ denotes the (i, j)-th element of $(A(f))^{-1}$. This result corresponds to the above-obtained (11) and, correspondingly to (13), we get

$$q_{ij}(f) = |(A(f))^{-1}_{ij}|^2 \sigma_j^2 \tag{22}$$

Thus, if the fitting of the multivariate autoregressive model of the form (18) is realized and the assumption of uncorrelatedness between $\varepsilon_i(s)$'s$(i=1, 2, ..., k)$ is accepted, then all the necessary quantities for the analysis described earlier in this section can be obtained by the calculations given in 3.2.2. (See, Program 5.3.3 for the actual calculations.)

To obtain the coefficients of (16) from (18) under the assumption that the $u_i(s)$'s, or $\varepsilon_i(s)$'s are uncorrelated, following calculations must be performed. First, put

$$c_i(l) = A_{ii}(l) \quad (l=1, 2, \cdots\cdots, M),\tag{23}$$

where M is the M of the autoregressive model of (18). Similarly, for $i \neq j$, put

$$A_{ijm} = A_{ij}(m) \quad (m=1, 2, \cdots\cdots, M).\tag{24}$$

$c_i(l)$ or A_{ijm} with l or m greater than M are put equal to zero. With this convention the $a_{ij}(m)$'s are given by

$$a_{ij}(1) = A_{ij1}$$
$$a_{ij}(m) = A_{ijm} + \sum_{l=1}^{m-1} c_i(l)\, a_{ij}(m-l) \quad (m=2, 3, 4, \cdots\cdots).\tag{25}$$

In this case, there is no upper limit to m and the values of $a_{ij}(m)$ may be computed for any range of m as required. By putting $A_{ij}(m)=0$ for $m>M$, (25) can be rewritten as

$$a_{ij}(1) = A_{ij}(1)$$
$$a_{ij}(m) = A_{ij}(m) + \sum_{l=1}^{m-1} A_{ii}(l)\, a_{ij}(m-l) \quad (m=2, 3, 4, \cdots\cdots).\tag{26}$$

Thus we can see that, under the assumption of the mutual uncorrelatedness of $u_i(s)$'s, the characteristics of the system and the noise sources can be obtained by (23) and (26) from the autoregressive model of form (18).

The fitting of the autoregressive model (18) can be realized almost automatically, including the determination of M, by the method explained in 3.2.2. If it is found that the assumption of the uncorrelatedness of the $\varepsilon_i(s)$'s is acceptable, i.e., if the coefficient of correlation between $\varepsilon_i(s)$ and $\varepsilon_j(s)$ is small for all $i \neq j$, we may examine the resulting estimate d_M of Σ and apply relation (25) to obtain an estimate of $\{a_{ij}(m)\}$ (Program 5.3.4) and use (21) and (22) to obtain the relative power contribution (Program 5.3.3). Thus we have a completely practical procedure for the application of the idea discussed in this section [13].

Here let us consider the assumption that the $u_i(s)$'s are uncorrelated. If there exists a significant noise source outside the group of observed variables $x_j(s)(j=1, 2, ..., k)$ that affects several $x_j(s)$'s, then naturally the corresponding $u_i(s)$'s, or the $\varepsilon_i(s)$'s, will show significant correlations. This represents the case where the available observations are insufficient for the analysis of the behavior of the system. Also in the case of a continuously varying system observed by sampling at constant time intervals, the correlations may take significant values when the sampling interval is too large. These observations suggest the necessity of careful examination of d_M, the estimate of Σ obtained by fitting the multivariate autoregressive model. It is possible, for example, under the assumption of Gaussianity, to test the uncorrelatedness of the $\varepsilon_i(s)$'s by using d_M. However, as N becomes large, any correlation will become statistically significant, no matter how small it is. Thus, in the present situation, we should rather check whether the values of the correlations can be ignored, in practice.

3.4   Some Key Points of Data Processing

In this section, we consider two problems related to data handling in time series analysis. One is concerned with the observation noise and the other is concerned with the sampling interval of observations. First we will consider the choice of the sampling interval.

3.4.1   Determination of the sampling interval

In 3.1.4 we already discussed the effect on the spectrum of sampling at a certain time interval Δt. Briefly, the spectral density of the sampled data can be obtained by cutting the original spectrum vertically into strips of constant width $1/\Delta t$, with the center strip centered at frequency 0, and then piling them up on the center strip. From this we can see that, if Δt is taken to be excessively large with coarse sampling some information on the original spectrum can be lost.

On the other hand, too small a Δt can also cause difficulties. One of them is the increase of the amount of data taken per unit time. This implies a severe requirement on the speed of the computer at the analysis and the resulting control stages. From this point of view, it is desirable to choose Δt as large as possible.

An even more serious problem is the effect of statistical errors when the model is fitted to the data in the time domain. The estimation of the power spectral density by the time domain fitting of an autoregressive model as described in 3.2.1, is known to minimize the integral over the frequency domain of the mean square of the ratio (estimated $-$ true)/true. The power spectrum of an ordinary signal usually takes low values at sufficiently high frequencies. Thus, when Δt is taken excessively small, values of the power spectrum at higher frequencies will, relatively speaking, be affected significantly by the effect of sampling and observation noise. When Δt is reduced still further, the disturbed range of the spectrum will become still wider and the corresponding autoregressive model will tend to respond excessively to the meaningless details of this widened spectrum. Thus it is quite risky to fit a model in the time domain to data taken with an excessively small Δt. In fact, in the examples hitherto published of analysis in the time domain, we sometimes found incomprehensible results when too small a sampling interval was adopted.

Summarizing these observations, we may conclude that we should take Δt as large as possible, so long as the corresponding power spectrum provides a reasonable description of the main portion of the true spectrum. Needless to say, the final decision must be made by trying various possible values of Δt for each particular application.

3.4.2   Treatment of observation noise

Generally speaking, any observation must be considered to be contaminated with some kind of noise. However, if the level of this noise is sufficiently low compared with that of the main signal, we may ignore the existence of the noise. In particular, in the case of fitting a statistical model, it is necessary to evaluate the influence of the noise in relation to the magnitude of the statistical error of the fitted model. When there is a noise which produces a significant effect on the fitting of the statistical model, we must develop some measure to reduce it to an insignificant level. To this end, an understanding of the statistical property of the noise is essential.

For example, for the measurement of the temperature of the gas inside a rotating kiln, the signal from the thermometer is sent out through a slip ring. In this case, we can expect that the noise due to the imperfect contact of the collector shoe that picks up the signal from the slip ring will be of the impulsive type. Since the original signal of the temperature will change slowly, if observations are taken with a sufficiently short sampling interval, contamination by noise should be detected by the occurrence of an extraordinarily large difference of a pair of adjacent data values. Further, if the frequency of occurrences of impulsive noise is not high, then the simple procedure of replacing the outlier by the average of its two adjacent values will be sufficiently useful for many practical applications.

When the observations are used directly for control, we must use a filter that uses only the past observations to realize the noise suppression. Apart from the simple observation noise caused within the measurement system, there are usually various intrinsic noise sources within a real system. Any signal without any significance for the purpose of control of the system under consideration appears to us as a noise. To handle such a noise property, an understanding of the physical characteristics of the noise is required. The design of a filter for the reduction of noise will naturally be based on the characteristic of each noise and will often be nonlinear, or linear but time varying. Practical examples are given in the construction of W_2 of 4.4 and W of 4.2.

The design of such a filter is highly empirical and there is no universal design procedure which is generally applicable to this type of problem. However, the common sense approach discussed in this section will often be quite useful. Hereafter we will simply ignore the influence of observation noise by assuming that it is relatively small. However, in a practical application sufficient attention must be paid to this point.

3.5 Design of a Statistical Control System

In this section we discuss the method of designing a control system, assuming that the analysis of the characteristics of the system has been successfully completed with the aid of the various procedures described in this book. This control system is based on the autoregressive modeling of multivariate time series and is designed according to the general theory of the optimal multivariate linear control system based on a quadratic loss function and the state space representation of the system.

The design procedure has the following features:

1) The determination of the basic model by real data is very simply performed by the method of autoregressive model fitting discussed in 3.2.

2) The control system can easily be designed by using a digital computer. The choice of the criterion function can be based on the result of the autoregressive model fitting and the design can easily be realized with a computer by using a dynamic programming procedure.

These features suggest that the method has already passed its feasibility study stage and is supported firmly both by a fully developed statistical information processing technique and by the efficient use of the electronic computer. Hence the method is extremely practicable.

In this section we will explain the method and briefly discuss the possibility of its further improvement.

3.5.1 State space representation of a dynamic system

The *state space representation* provides the simplest description of the behavior of a dynamic system. The *state* represents the minimum necessary information of the present and past of the system required to describe its future behavior. This means that the state is defined in such a way that once the present state has been determined by the present and past behavior of the system, this knowledge alone is sufficient to optimally predict the behavior of the system in the future. If this concept of state is accepted the dynamics or temporal pattern of the behavior of the system can be described precisely as the law that governs the transition of the state with time. System variables which are directly observed are generally defined as functions of the state.

We now consider a dynamic system with discrete time parameter $s=...,$ $-1, 0, 1, ...$ Let $y(s)$, $z(s)$ and $x(s)$ denote the input to the system, the state, and the observed output, respectively. Then, from the preceding discussion, the representation of a linear dynamic system is given by

$$z(s+1)=\Phi(s+1, s)\, z(s)+G(s)\, y(s)$$
$$x(s)=H(s)\, z(s) \tag{1}$$

where, $z(s)$, $y(s)$, $x(s)$ denote column vectors and $\Phi(s+1, s)$, $G(s)$, $H(s)$ denote the corresponding matrices with appropriate dimensions. In particular, $\Phi(s+1, s)$ is called the *transition matrix* that controls the transition of the state $z(s)$ at time s to the state $z(s+1)$ at time $s+1$. If the whole system is time invariant, $\Phi(s+1, s)$, $G(s)$ and $H(s)$ reduce to the constant matrices Φ, G and H, respectively, and representation (1) reduces to

$$z(s+1)=\Phi z(s)+Gy(s)$$
$$x(s)=Hz(s). \tag{2}$$

As is clear from this representation, the state $z(s)$ preserves all the information about the effect on the system produced by the past input $y(s-1)$, $y(s-2)$, ...

The most significant use of this representation is the convenience that the representation provides for the design of a control system. Once such a representation is given, even an extremely complex multivariate system can formally be handled as easily as a univariate system. The complexity of the system is then simply represented by increasing the dimensions of the matrices Φ, G, H and the corresponding vectors. What would happen if the concept of state is extended to a multivariate statistical dynamic system? This problem will be discussed in the next subsection.

From the point of view of control theory, there are well-known problems of controllability and observability that discuss the controllability of the system through the input $y(s)$ and the determination of the state $z(s)$ through the observation of $x(s)$, respectively. However, these concepts are related to the redundancy of the representation and with a model fitting procedure that eliminates redundant models the resulting statistically identified model is expected to satisfy these conditions automatically. Thus we will not be concerned with those problems here.

3.5.2 State space representation of a stationary time series

What would be the state space representation of a stationary time series $x(s)$? By extending the concept of state discussed in 3.5.1 to this case, we can see that the state $z(s)$ must contain all the information about the future behavior of $x(s)$ produced by the present and past values of $x(s)$. Note that here the information is restricted to that which is linearly related to the future behavior of $x(s)$ in the sense of mean square error. In this case, corresponding to representation (2) of 3.5.1, we might consider the representation

$$z(s+1)=\Phi z(s)+u(s)$$
$$x(s)=Hz(s)$$

(1)

where $u(s)$ denotes the part of $z(s+1)$ that is statistically linearly unrelated to the behavior of $z(s)$. As the state $z(s)$ is expected to represent the information from the past of the system exhaustively, $u(s)$ must be uncorrelated with $z(s)$ and its past.

For simplicity, we will assume that every variable has mean zero. Now, since $u(s+1)$ is uncorrelated with $z(s+1)$ and its past, it must be uncorrelated with $u(s)$ and its past and thus $u(s)$ is a multivariate white noise. Accordingly, it is natural to consider representation (1) defined with a white noise $u(s)$ that is uncorrelated with $z(s)$ and its past, as the state space representation of a stationary time series. By replacing the concept of uncorrelatedness by independence, we obtain from (1) a representation in a stronger sense. In the case of a Gaussian process there is no difference between these two representations. It is tacitly assumed that the related variables have finite second order moments.

Now, what are the motives for the consideration of the state space representation of a stationary time series? Needless to say, convenience of control system design, which is to be discussed shortly, is one motive. Another is the possibility of implementing the principle of parsimony discussed in 1.5 by pursuing the simplest representation of time series based on the concept of state. However this latter objective is not carried out completely in this book. Here, we only intend to follow the first motive, and try to derive a state space representation from the autoregressive model. If this is done, we can obtain a state space representation immediately from a given set of data, as we already have a completely practical procedure for the fitting of an autoregressive model.

By (1) of 3.2.2 the autoregressive representation of a stationary time series $X(s)$ is given by

$$X(s)=\sum_{m=1}^{M} A(m)\, X(s-m)+U(s)\,.$$

(2)

Here, $U(s)$ denotes a white noise uncorrelated with the past values of $X(s)$. Now we consider the time point $s-1$, and assume that the past values $X(s-1)$,

$X(s-2)$, ... are given. The information from the past that is linearly related with the future values $X(s)$, $X(s+1)$, ... is, from (2), completely exhausted by $X(s-1)$, $X(s-2)$... $X(s-M)$. This observation leads us immediately to the following state space representation. Let,

$$Z(s) = \begin{bmatrix} X(s) \\ X(s-1) \\ \vdots \\ X(s-M+1) \end{bmatrix} \tag{3}$$

then

$$Z(s) = \Psi_0 Z(s-1) + V_0(s)$$
$$X(s) = H_0 Z(s) \tag{4}$$

where

$$\Psi_0 = \begin{bmatrix} A(1) & A(2) & A(3) & \cdots\cdots & A(M-1) & A(M) \\ I & O & O & \cdots\cdots & O & O \\ O & I & O & \cdots\cdots & O & O \\ \vdots & \vdots & \vdots & & \vdots & \vdots \\ O & O & O & \cdots\cdots & I & O \end{bmatrix}$$

$$V_0(s) = \begin{bmatrix} U(s) \\ O \\ \vdots \\ O \end{bmatrix}, \qquad H_0 = [I \quad O \quad O \quad \cdots\cdots \quad O]$$

where I and O denote the unit and zero matrix (or zero vector of appropriate dimensions), respectively.

The representation (4) is not unique. To predict $X(s)$ from $Z(s-1)$ by using (4), we must always compute

$$\sum_{m=1}^{M} A(m) X(s-m).$$

If we follow (4) formally, we must perform all the necessary computations when we obtain a new observation at time $s-1$. This turns out to be disadvantageous when an on-line operation is required, as it takes time to obtain a new prediction by this procedure. Consequently, a representation giving the minimum amount of computation for prediction is required; see, for example [14]. We will not pursue this point to the limit. Instead, we consider a new state obtained from the present and past data by transforming them into the terms that are to be used directly in the definition of future predictors. By this representation, we will be able to obtain the prediction of $X(s)$ only by performing the computation related with the latest observation $X(s-1)$. Other calculations of updated values necessary for the next step can be done during the time before we get the next observation $X(s)$. If we replace s of (2) by $s+l$, we get

$$X(s+l) = \sum_{m=l+1}^{M} A(m) \, X(s+l-m) + \sum_{m=1}^{l} A(m) \, X(s+l-m) + U(s+l)$$

$$= \sum_{i=1}^{M-l} A(l+i) \, X(s-i) + \sum_{j=0}^{l-1} A(l-j) X(s+j) + U(s+l) . \qquad (5)$$

Accordingly, if we define

$$Z_l(s) = \sum_{i=1}^{M-l} A(l+i) \, X(s-i) \quad (l=0, 1, \cdots, M-1) \qquad (6)$$

this represents that part of the predictor of $X(s+l)$ that depends only on the observations $X(s-1)$, $X(s-2)$, ... obtained up to time $s-1$, and in particular $Z_0(s)$ gives the prediction of $X(s)$ at time $s-1$. Recall the relations

$$Z_l(s) = \sum_{j=1}^{M-l-1} A(l+1+j) \, X(s-1-j) + A(l+1) \, X(s-1)$$

$$= Z_{l+1}(s-1) + A(l+1) \, X(s-1) \quad (l=1, \cdots, M-2) \qquad (7)$$
$$Z_{M-1}(s) = A(M) \, X(s-1) .$$

Consider the moment when a new observation $X(s)$ is obtained. Then $Z_0(s)$ is redefined by

$$Z_0(s) = X(s) , \qquad (8)$$

and the following equations hold

$$Z_0(s) = Z_1(s-1) + A(1) \, Z_0(s-1) + U(s)$$
$$Z_l(s) = Z_{l+1}(s-1) + A(l+1) \, Z_0(s-1) \quad (l=1, 2, \cdots, M-2) \qquad (9)$$

$$Z_{M-1}(s) = \qquad\qquad A(M)\, Z_0(s-1).$$

Accordingly, if we redefine $Z(s)$ by

$$Z(s) = \begin{bmatrix} Z_0(s) \\ Z_1(s) \\ \vdots \\ Z_{M-1}(s) \end{bmatrix} \qquad\qquad (10)$$

(4) can be written as

$$Z(s) = \Phi_0\, Z(s-1) + V_0(s)$$
$$X(s) = H_0\, Z(s) \qquad\qquad (11)$$

where

$$\Phi_0 = \begin{bmatrix} A(1) & I & O & \cdots\cdots & O \\ A(2) & O & I & \cdots\cdots & O \\ \vdots & & \vdots & \vdots & \vdots \\ A(M-1) & O & O & \cdots\cdots & I \\ A(M) & O & O & \cdots\cdots & O \end{bmatrix} \qquad V_0(s) = \begin{bmatrix} U(s) \\ O \\ \vdots \\ O \\ O \end{bmatrix} \qquad (12)$$

$$H_0 = [\, I \quad O \quad O \quad \cdots\cdots \quad O\,].$$

From (11) we can see that we may use $\Phi_0 Z(s-1)$, obtained by putting $V_0(s) = O$ in (11), as the predictor of $Z(s)$ at time $s-1$ and thus the predicted value $\hat{X}(s)$ of $X(s) = Z_0(s)$ can be obtained, when $X(s-1)$ is observed, by the first equation of (9) as

$$\hat{X}(s) = Z_1(s-1) + A(1)\, X(s-1) \qquad\qquad (13)$$

which requires only a small amount of computation. Thus we can see that (11) and (12) give a natural representation from the point of view of prediction. This representation is also suitable for the design of a control system which is described in the following discussion.

From the point of view of control, the vector of observations $X(s)$ can be partitioned into two groups. The first part, represented by the vector $x(s)$ represents the output of the system itself and other variables which are non-manipulable and the second part denoted by $y(s)$ represents the inputs to the system that can be manipulated. Assume that $x(s)$ and $y(s)$ are r- and l-dimensional column vectors respectively, and that

$$X(s) = \begin{matrix} \overset{\leftarrow 1 \rightarrow}{\begin{bmatrix} x(s) \\ y(s) \end{bmatrix}} \end{matrix}. \qquad (14)$$

We call $x(s)$ the *controlled variable*, and $y(s)$ the *manipulated variable*. Assume further that at the time of data acquisition $y(s)$ has its own noise source. This assumption usually holds, as for example when $y(s)$ is the manipulated variable of a system controlled by a human operator. On the other hand, when $y(s)$ is the output of a linear feedback loop that is defined by a computer program which connects the output of the system to the input, the fitting of the autoregressive model by the procedure described in the preceding chapter is inapplicable. This point will be discussed again later.

Even if relation (11) is obtained from the observed data, when control is contemplated it is meaningless to consider the prediction of the behavior of $y(s)$ by (11), as $y(s)$ is now a vector that is going to be manipulated. This shows that only the first r components of the $(r+l)$-vector $Z_i(s)$ of (10) are related to the forecasting of the controlled variable $x(s)$. Thus we redefine this portion of the vector by $z_{i,s}$. That is,

$$Z_i(s) = \begin{matrix} \overset{\leftarrow 1 \rightarrow}{\begin{bmatrix} z_{i,s} \\ * \end{bmatrix}} \end{matrix} \qquad (i=0,1,2,\cdots\cdots, M-1), \qquad (15)$$

where $*$ denotes the part no longer required. Since it holds that

$$Z_0(s) = \begin{matrix} \overset{\leftarrow 1 \rightarrow}{\begin{bmatrix} x(s) \\ y(s) \end{bmatrix}} \end{matrix} \qquad (16)$$

we have

$$z_{0,s} = x(s). \qquad (17)$$

Thus, by putting

$$Z_s = \begin{bmatrix} z_{0,s} \\ z_{1,s} \\ \vdots \\ z_{M-1,s} \end{bmatrix} \tag{18}$$

(11) and (12) are transformed into

$$Z_s = \Phi\, Z_{s-1} + \Gamma\, Y_{s-1} + W_s$$
$$x(s) = H\, Z_s$$

$$\Phi = \begin{bmatrix} a_1 & I & O & \cdots & O \\ a_2 & O & I & \cdots & O \\ \vdots & \vdots & \vdots & & \vdots \\ a_{M-1} & O & O & \cdots & I \\ a_M & O & O & \cdots & O \end{bmatrix}, \qquad \Gamma = \begin{bmatrix} b_1 \\ b_2 \\ \vdots \\ b_{M-1} \\ b_M \end{bmatrix} \tag{19}$$

$$W_s = \begin{bmatrix} u(s) \\ O \\ \vdots \\ O \\ O \end{bmatrix}, \qquad H = [\, I \quad O \quad \cdots \quad O \quad O\,], \qquad Y_s = y(s)$$

where a_m, b_m, $u(s)$ etc. are obtained from (12) by the relations

$$A(m) = \begin{bmatrix} a_m & b_m \\ * & * \end{bmatrix}, \qquad U(s) = \begin{bmatrix} u(s) \\ * \end{bmatrix}. \tag{20}$$

This result shows that the only part used for the design of the control system is the section consisting of the upper r rows of the autoregressive coefficient matrices $A(m)$.

Representation (19) takes the form of (1) of 3.5.1 with respect to $y(s)$

which is the input to the system and it takes the form of (1) of this subsection with respect to the random input W_s. As will be discussed in the next section, with the help of this formula (20), we can easily proceed to the design of the control system.

The preceding discussion explains that, at least in principle, it is possible to identify representation (19) of the system with a record of stationary operation by a human operator, as in the case of the example of a cement kiln. We already know a practical procedure for the fitting an autoregressive model to a stationary time series. This means that the system representation (19) can always be determined in a practical application.

3.5.3 Optimum controller design under a quadratic criterion

In this section, we consider the system given by (19) of the preceding subsection 3.5.2 and for the evaluation of the control input $Y_s(=y(s))$ we adopt the *quadratic criterion* defined by

$$J_I = E\{K_I\}$$
$$K_I = \sum_{s=1}^{I} \{Z_s' \, Q(s) \, Z_s + Y_{s-1}' \, R(s) \, Y_{s-1}\} \tag{1}$$

where I is a properly chosen positive integer. Here $Q(s)$ and $R(s)$ denote $Mr \times Mr$ and $l \times l$ non-negative definite matrix, respectively. In particular, we will assume that $R(s)$ is positive definite. Z_s' and Y_{s-1}' denote row vectors obtained by transposing Z_s and Y_{s-1}, respectively, and E denotes mathematical expectation.

Where Z_0 is given, K_I can be given as a composition of the effect of the sequence of the inputs $Y_0, Y_1, ..., Y_{I-1}$ and that of the noises $W_1, W_2 ..., W_I$. From (17), (15), (14), and (6) of 3.5.2, it can be seen that Z_0 is determined when $x(0), x(-1), y(-1), x(-2), y(-2), ..., x(-M+1), y(-M+1)$ are given. The first and second terms of the right hand side of the second, K_I defining, equation of (1) take non-negative values and represent the effect of the deviation of the state of the system from the null state $Z_s = O$(zero vector) and the loss due to the variation of the manipulated variable. For a given period of operation of length I, the sequence of Y_s's that minimizes J_I defines the optimal control. When we are only interested in the use of a quantity obtained by an appropriate linear transform KZ_s of the state Z_s, where K denotes an appropriate matrix, the term $Z_s Q(s) Z_s$ on the right hand side of K_I can be rewritten as

$$Z_s' \, K' \, Q_K(s) \, K Z_s \,, \tag{2}$$

where $Q_K(s)$ is a properly chosen matrix. This result is equivalent to putting $Q(s)=K'Q_k(s)K$ in (1). For example, when we pay our attention only to the behavior of $x(s)=HZ_s$ and the evaluation is based on an $r \times r$ non-negative matrix $Q_1(s)$, the matrix $Q(s)$ of (1) takes the formon

$$Q(s) = H' Q_1(s) H$$

$$= \begin{array}{c} \\ \updownarrow r \\ \\ {}_{(M-1)r} \updownarrow \end{array} \overset{\leftarrow r \rightarrow \; \leftarrow (M-1)\, r \rightarrow}{\begin{bmatrix} Q_1(s) & O \\ O & O \end{bmatrix}}, \qquad (3)$$

an where O denotes a zero matrix. The application to be discussed in this book is an example of this particular form.

To obtain the sequence $Y_s(s=0, 1, ..., M-1)$ that minimizes J_I, we apply the well-known optimality principle of dynamic programming. Since we have

$$K_I = Z_I' Q(I) Z_I + Y_{I-1}' R(I) Y_{I-1} + K_{I-1} , \qquad (4)$$

by taking the expectations of the quantities on both sides of the equation we get

$$J_I = E\{Z_I' Q(I) Z_I + Y_{I-1}' R(I) Y_{I-1}\} + J_{I-1} . \qquad (5)$$

From (19) of 3.5.2 we have

$$Z_I = \Phi Z_{I-1} + \Gamma Y_{I-1} + W_I , \qquad (6)$$

and, since W_I is uncorrelated with the other variables, it holds that

$$E Z_I' Q(I) Z_I = E W_I' Q(I) W_I$$
$$+ E(Y_{I-1}' \Gamma' + Z_{I-1}' \Phi') Q(I) (\Phi Z_{I-1} + \Gamma Y_{I-1}). \qquad (7)$$

As the first term on the right hand side of (7) is a constant which is independent of Y_{I-1}, we have only to consider the sum of the second term and $E(Y'_{I-1} R Y_{I-1})$ to determine Y_{I-1}. To minimize this sum, we have only to find Y_{I-1} that minimizes

$$(Y_{I-1}' \Gamma' + Z_{I-1}' \Phi') Q(I) (\Phi Z_{I-1} + \Gamma Y_{I-1}) + Y_{I-1}' R(I) Y_{I-1} . \qquad (8)$$

Since (8) can be rewritten as

$$Z_{I-1}' \, \Phi' \, Q(I) \, \Phi \, Z_{I-1} + Z_{I-1}' \, \Phi' \, Q(I) \, \Gamma \, Y_{I-1}$$
$$+ Y_{I-1}' \, \Gamma' \, Q(I) \, \Phi \, Z_{I-1} + Y_{I-1}' \, (R(I) + \Gamma' \, Q(I) \, \Gamma') \, Y_{I-1} \, , \qquad (9)$$

the control input Y_{I-1} that minimizes (8) can be obtained by putting the differential coefficient of (9) with respect to each component of Y_{I-1} equal to 0, i.e., by solving the equation

$$(R(I) + \Gamma' \, Q(I) \, \Gamma') \, Y_{I-1} + \Gamma' \, Q(I) \, \Phi \, Z_{I-1} = 0 \, . \qquad (10)$$

The desired Y_{I-1} is then given by

$$Y_{I-1} = - \, (R(I) + \Gamma' \, Q(I) \, \Gamma')^{-1} \, \Gamma' \, Q(I) \, \Phi \, Z_{I-1} \qquad (11)$$

which is a linear transform of Z_{I-1}. This is a natural result, as the state Z_{I-1} constains all the necessary information about the future behavior of the system that is statistically linearly related with the past behavior.

Substituting Y_{I-1} of (9) by (11), we obtain the minimum value of (8) as

$$Z_{I-1}' \, \Phi' \, (Q(I) - Q(I) \, \Gamma \, (R(I) + \Gamma' \, Q(I) \, \Gamma')^{-1} \, \Gamma' Q(I)) \, \Phi \, Z_{I-1} \, . \qquad (12)$$

Accordingly if we put

$$M(I-1) = Q(I) - Q(I) \, \Gamma \, (R(I) + \Gamma' Q(I) \Gamma')^{-1} \Gamma' Q(I)$$
$$S(I-1) = \Phi' \, M(I-1) \, \Phi \qquad (13)$$

(12) is given by

$$Z_{I-1}' \, S(I-1) \, Z_{I-1} \, . \qquad (14)$$

Thus, if we define Y_{I-1} by (11), we get

$$J_I = E \, W_I' \, Q(I) \, W_I + E \{ Z_{I-1}' \, S(I-1) \, Z_{I-1} + K_{I-1} \} \, . \qquad (15)$$

Now, let $\tilde{K}_{I-1}$ denote the term inside the braces {} in (15). Then, since we have, corresponding to (4),

$$K_{I-1} = Z_{I-1}' \, Q(I-1) \, Z_{I-1} + Y_{I-2}' \, R(I-1) \, Y_{I-2} + K_{I-2} \qquad (16)$$

we get the relation

$$\tilde{K}_{I-1} = Z_{I-1}' \, \tilde{Q}(I-1) \, Z_{I-1} + Y_{I-2}' \, R(I-1) \, Y_{I-2} + K_{I-2} \, , \qquad (17)$$

where

$$\tilde{Q}(I-1) = S(I-1) + Q(I-1) . \qquad (18)$$

From this result, it is obvious that we can obtain the optimal control input Y_{I-2} by replacing Q by $\tilde{Q}$ and I by $I-1$ in (11). Similarly, to obtain $Q(I-2)$ which is necessary for the calculation of Y_{I-3}, we have only to replace Q by $\tilde{Q}$ and I by $I-1$ in (13) and also replace I of (18) by $I-1$.

Thus by defining $\tilde{Q}(I)$ by

$$\tilde{Q}(I) = Q(I) \qquad (19)$$

we can determine Y_{I-i} recursively for $i=1, 2, ..., I$ by

$$Y_{I-i} = -(R(I-i+1) + \Gamma' \tilde{Q}(I-i+1) \, \Gamma)^{-1} \, \Gamma' \, \tilde{Q}(I-i+1) \, \Phi \, Z_{I-i}$$
$$M(I-i) = \tilde{Q}(I-i+1) - \tilde{Q}(I-i+1)\Gamma(R(I-i+1)$$
$$\qquad\qquad\qquad + \Gamma'\tilde{Q}(I-i+1)\Gamma)^{-1}\Gamma'\tilde{Q}(I-i+1) \qquad (20)$$
$$\tilde{Q}(I-i) = \Phi' \, M(I-i) \, \Phi + Q(I-i) .$$

Now we consider a stationary system and put

$$Q(s) = Q$$
$$R(s) = R \qquad (21)$$

and define P_i and M_i by

$$P_i = \tilde{Q}(I-i)$$
$$M_i = M(I-i) . \qquad (22)$$

Then by starting from

$$P_0 = Q \qquad (23)$$

and calculating recursively

$$M_i = P_{i-1} - P_{i-1}\Gamma \, (R + \Gamma' P_{i-1}\Gamma)^{-1} \, \Gamma' \, P_{i-1}$$
$$P_i = \Phi' \, M_i \Phi' + Q \qquad (24)$$

we obtain the gain matrix G_i, that is the coefficient of Z_{I-i} of the right hand side of the first equation of (20), by

$$G_i = -(R + \Gamma' P_{i-1} \Gamma)^{-1} \Gamma' P_{i-1} \Phi .$$ (25)

Thus the optimal control input is given by

$$Y_{I-i} = G_i Z_{I-i} .$$ (26)

Now consider the gain matrix corresponding to $i = I$. This is given by

$$G_I = -(R + \Gamma' P_{I-1} \Gamma)^{-1} \Gamma' P_{I-1} \Phi$$ (27)

and defines the gain for the feedback loop that generates the optimal input at time $s = 1$ that takes into consideration the behavior of the system in the period $s = 1$ to I.

If we apply the feedback control defined by putting $Y_s = G_I Z_s$ with G_I fixed, this always gives the first input of a control that is designed to optimize the performance looking I periods ahead from the present time. Thus when I is increased sufficiently, G_I will converge to a constant matrix G. This G will give a stationary control that will be optimal for infinitely large I. This shows that a practically useful feedback control will be defined by

$$Y_s = G Z_s .$$ (28)

The above calculation of G_I can be realized by Program 5.5.1 OPTDES, where Q is restricted to the form of (3) where $Q_1(s)$ is fixed to be Q_1, a constant matrix. In the calculation of G_I, care must be taken not to use an unnecessarily large I, as rounding errors will accumulate quickly. It is effective to compare the result with the corresponding result obtained by computation with a higher precision. Further characteristics of a control system thus obtained must be checked by various simulation experiments to ensure its practical applicability.

The preceding design of the control system is based on the model of the system obtained by fitting a multivariate autoregressive model to a record of observations where each variable has its own noise source. Generally, relations between variables identified by a statistical method provide merely a phenomenal description of mutual relations between the variables. When such a description is going to be used for the practical purpose of control, it is desirable and often necessary to check by partial experiments or by careful observations of the sytem whether the effect expected from the fitted model is actually obtainable by the adjustment of each manipulated variable.

For a system with a feedback defined by (28), the first equation of (19) of 3.5.2 is replaced by

$$Z_s = (\Phi + \Gamma G)\, Z_{s-1} + W_s \; . \tag{29}$$

The value of Z_s obtained by letting $W_s = 0$ is given by

$$\hat{Z}_s = (\Phi + \Gamma G)\, Z_{s-1} \tag{30}$$

and provides the prediction of Z_s based on the previous state of the system Z_{s-1}.

Actually, it is the vector $\hat{z}_{0,s}$ of the first r components of this $\hat{Z}_s$ that defines the predictor of the controlled variable $z_{0,s}$ of the system at time s. The remaining components of $\hat{Z}_s$ are identical to the corresponding components of Z_s. Thus we can obtain Z_s simply by replacing $\hat{z}_{0,s}$ by the observed value of the controlled variable $x(s)$. The difference between this $z_{0,s} = x(s)$ and its predictor $\hat{z}_{0,s}$ is the white noise $u(s)$. Thus, if the correct model is used, the prediction errors should form a white noise. When some change occurs within the system, that will cause a systematic deviation of the forecasting error from the whiteness. The practical implication of this will be discussed in section 4.5.

3.5.4   Selection of the performance criterion

The first problem in the application of the present design procedure is the choice of the matrices Q and R. A practical approach to this problem is to check the performance of the system obtained by a particular pair, Q and R, and then adjust Q or R until a final decision is reached. (See Program 5.5.2.)

In this approach, the most difficult part is to make a fairly good initial choice of Q and R. Once a reasonable choice is made further adjustments are relatively easy. The characteristics of the present model suggest that the procedure to be described below will provide a good initial guess. In fact, the procedure was found to be particularly useful in the case of the controller design of the cement kiln.

First, we assume that the original data represent the normal behavior of the system. In this case, the variance, or the mean squared deviation of the observations from the mean, of a manipulated variable provide an initial guess of the realistic range of the variation of the manipulated variable. This and any other considerations will lead to the final determination of the allowable range of the variation of the manipulated variable. When a control system is implemented, the variance of the manipulated variable must, of course, remain within this allowable range. In practice, the variation of a manipulated variable is restricted to a certain fixed finite range and we try to keep the probability of hitting the boundary sufficiently small. This adjustment can be realized by modifying R. For example, if R is taken to be a positive diagonal matrix, the weighted sum of the variances of the manipulated variables defines the performance criterion. In this case, we can obtain a design which reduces the variances of some of the manipulated variables by increasing the corresponding diagonal components of R.

For a general R, the variation of the i-th manipulated variable can be evaluated at a constant c times its real value, by multiplying the i-th column and the i-th row of R by c. For evaluation of the variance, use formula (19) of 3.5.2 and let

$$Y_{s-1} = G Z_{s-1}$$

$$Z_s = \Phi Z_{s-1} + \Gamma Y_{s-1} + W_s \ . \tag{1}$$

The simulation experiment realized by using random numbers to define W_s will provide a useful estimate of the variance. Necessary computations can be performed by using Program 5.5.2 OPTSIM and Program 5.5.3 WNOISE.

Obviously the evaluation of the variances can be performed analytically. For this purpose, we can use the relation

$$E Z_s Z_s' = (\Phi + \Gamma G) E Z_{s-1} Z_{s-1}' (\Phi + \Gamma G)' + E W_s W_s' \ . \tag{2}$$

Now let us consider the determination of Q. IF we restrict Q to be of the form (3) of 3.5.3 and put $Q_1(s) = Q_1$, the variation of the controlled variable is evaluated by Q_1. Further, when Q_1 is a positive diagonal matrix, the weighted sum of the variances of controlled variables constitutes a performance criterion. By increasing the value of the (i, i)-th component of Q_1, we obtain a design that will reduce the variance of the i-th controlled variable. Obviously the choice of the relative weights of variances of the controlled variables is the main problem here.

It can be seen by (1) that even with perfect control the variances of the controlled variables remain because of the existence of W_s, or in the formulation (19) of 3.5.2 because of the existence of $u(s)$. Thus it seems reasonable to evaluate the variances of the components of $x(s)$ relative to the magnitudes of the corresponding variances of the components of $u(s)$. Extending this idea to the general form of Q_1, we find that the inverse of the variance-covariance matrix of $u(s)$ wil be a reasonable choice for Q_1.

Summarizing these observations, we have the following procedure:

1) As Q_1, take the diagonal matrix with the diagonal elements defined by the reciprocals of the corresponding diagonal elements of the estimate of the variance-covariance matrix of $u(s)$. This estimate is given as the upper left $r \times r$ submatrix of the matrix d_M, obtained by fitting a multivariate AR model by procedure 3.2.2 to the record of observations

$X(s)$ given by (14) of 3.5.2. The choice of the order M is different from that of 3.2.2 and will be discussed in the following section. When the upper left $r \times r$ submatrix of d_M is significantly different from a diagonal matrix, use d_M^{-1} as Q_1.

2) As R, take the diagonal matrix with the diagonal elements set equal to the reciprocals of the allowable limits of the variances of the corresponding components of the manipulated variable $y(s)$.

3) With G obtained by using the above Q and R perform a simulation experiment using (1) for an appropriate length of time and examine the result.

4) Using the results of the simulation, compute the estimates of the variances of the components of $y(s)$. If, for some components, the estimated variance is above the allowable limit, decrease the value of the corresponding diagonal element of R. When the estimated variance is below the allowable limit decrease the corresponding diagonal element of R. This type of adjustment can be realized easily at the initial stage of the calculation, by multiplying the original component of R by the ratio of the estimated variance obtained from the simulation to its allowable limit.

5) Repeat 3) and 4) several times, using one and the same realization of the white noise $u(s)$ for the simulation. For the generation of a realization of $u(s)$, see 3.2.3. By the present procedure we can usually obtain a matrix G which is useful for practical applications.

6) If further adjustment is necessary modify Q and R as required.

3.5.5 Autoregressive model fitting for controller design

For multivariate autoregressive model fitting for the design of a control system, the procedure of 3.2.2 should be applied to

$$X(s) = \begin{bmatrix} x(s) \\ y(s) \end{bmatrix}, \tag{1}$$

where $x(s)$ denotes an r-dimensional vector of the controlled variables, and $y(s)$ an l-dimensional vector of the manipulated variables. When the estimate $A_M(m)$ of the coefficient matrix $A(m)$ of the autoregressive expression

$$X(s) = \sum_{m=1}^{M} A(m)\, X(s-m) + U(s) \tag{2}$$

is obtained, the model (19) of 3.5.2 is determined by putting

$$\begin{matrix} \overleftarrow{r} \;\; \overleftarrow{l} \\ \left. \begin{bmatrix} a_m & b_m \\ & \\ * & * \end{bmatrix} \right\} \end{matrix} = A_M(m) \quad (m=1, 2, \cdots\cdots, M) \qquad (3)$$

as in the case of (29) of 3.5.2.

The coefficients a_m and b_m are determined by data, and thus the corresponding model is an estimate of the true structure (19) of 3.5.2. Naturally the estimate contains stochastic error of which the expected magnitude depends on the length N of the data. The main problem here is the determination of the order M of the model. We adopt the following method which is different from that of 3.2.2 and is obtained by ignoring the modeling of the manipulated variable in the forecasting equation.

In place of MFPE(M) of 3.2.2, we use the following criterion

$$\text{FPEC } (M) = \left(1+\frac{Mk+1}{N}\right)^{r}\left(1-\frac{Mk+1}{N}\right)^{-r}\|d_{r,M}\| \quad (M=1, 2, \cdots\cdots, L) . \quad (4)$$

Here, $k=r+l$ and $d_{r,M}$ denotes the upper left $r \times r$ submatrix of d_M of 3.2.2 that provides an estimate of the variance covariance matrix of the innovations of the controlled variables. $\|d_{r,M}\|$ denotes the determinant of $d_{r,M}$. We adopt that M which gives the minimum of FPEC(M)($M=1, 2, ..., L$) as the order of the model. (See Program 5.3.2.)

In order to gain insight into the statistical error of the result thus obtained, perform the simulation experiment with the autoregressive model and apply the present procedure to the resulting data. Several repetitions of the simulation will give a fairly clear idea of the properties of the statistical error of the estimation procedure.

The foundation of the model fitting procedure discussed above is the basic assumption that the noise sources, or innovations, of the controlled and manipulated variables are mutually independent. Also the derivation of the order determination criterion is based on the assumption that the innovations $U(s)$ are mutually independent for different s. This last assumption holds when $X(s)$ is a Gaussian process. The former assumption means that there is no common factor that affects the controlled and manipulated variables simultaneously. In the application of the present method to a real problem, the choice of $x(s)$ and $y(s)$ itself often becomes a problem and the present assumption means that every important controlled variable that activate the feedback loop should be taken into account.

On the other hand, from the point of view of the expected accuracy of the fitted model, the unnecessary inclusion of any variables into the model merely reduces the accuracy. In particular, we must avoid including two or more variables that show very similar behavior, or a variable that can be represented as a linear combination of others. This is clear from the fact that essentially the inverse of the variance covariance matrix of $x(s-1), x(s-2), ...,$ $x(s-M)$ is used to determine $A_M(m)$. If variables of the types mentioned above are included, the inverse matrix will show a drastic change for a small change of the original matrix. This loss of the numerical stability naturally means a large estimation error.

In such a case we may calculate $\text{RFPEC}(M) = \text{FPEC}(M)/\text{FPEC}(0)(M=1,$ $2, ..., L)$ for various combinations of $x(s)$ and $y(s)$, and consider the use of the combination that gives the minimum of $\text{RFPEC}(M)$. This is, of course, only one possible course of action and generally the choice of the variables should depend heavily on prior knowledge of the structure of the system under consideration. However, when it is hard to make any reasonable judgement about the inclusion or exclusion of a controlled variable into or from the model the use of some reasonably defined mechanical procedure of the above type seems to be useful. Obviously the accuracy of the decision depends on the length of data available.

The preceding discussion is concerned with the so-called "principle of parsimony" and $\text{MFPE}(M)$ or $\text{FPEC}(M)$ is used as a device to realize this principle. These criteria provide estimates of the deviation of the probability distribution determined by the true model from the fitted models. The measure of the deviation is related to the so-called information number when the distributions are assumed to be Gaussian.

The complexity of the model to be adopted depends on the information contained in the available data and neither an excessively simplified model nor an unnecessarily complex model is appropriate for practical use. This point should be kept in mind whenever the determination of the model is performed on the basis of a record of finite length. In the conventional approach to model fitting this point has not been formulated explicitly, which has caused serious difficulty in practical applications. The basic idea of the previously discussed determination procedure for the order M of a model was first introduced by one of the present authors and, as far as our experience is concerned, its practical utility is quite remarkable.

3.5.6   The direction of future developments

We have discussed the use of the state space representation of a stochastic system obtained by fitting an autoregressive model. From the point of view of controller design and also from that of the "principle of parsimony" of model selection, the state space representation with a state of minimal dimension is preferred. The subject of the minimal representation of a system has been pursued by R. E. Kalman mainly from the point of view of the design of a system; see, for example, theorem 1 of page 224 of [15]. Our procedure based on MFPE(M) or FPEC(M) can be considered to be a realization of the "principle of parsimony" when we model the system based on real data.

There is a problem that relates to the linking of these two types of research. This problem is the realization of an efficient modeling procedure that will possibly require fewer coefficients than the autoregressive model. On this point, we can proceed one step further by using the mixed *autoregressive moving average model*, defined by an autoregressive model with white noise replaced by a weighted sum of lags of the white noise. However, the difficulty of numerical calculation required for fitting such a model to real data is quite significant and, compared with the fitting of an autoregressive model which can be realized by a very simple calculation, the fitting of the mixed type model of a multivariate time series must be considered to be still in the experimental stage.

Even in the case of the autoregressive model, it is possible to increase the accuracy of the fitted model further by putting unnecessary coefficients equal to zero. However, considering the complexity of the required calculation, we believe that the use of an autoregressive model in the form treated in this book has definite advantage.

Hitherto, the difficulty of establishing the correspondence between a theoretical model and the actual complex phenomenon has often been ignored and this made the application of the results of theoretical studies impractical. In this sense, the practical application of the multivariate autoregressive model provides an exceptional example that overcomes the difficulty. We hope that similar useful statistical models will be developed that can easily be applied for the purpose of the analysis and control of a complex system. Once an appropriate numerical procedure is developed, the use of the above mentioned mixed autoregressive moving average model will also easily be put into practice.

REFERENCES

[1]   G. M. Jenkins and D. G. Watts, *Spectral Analysis and Its Applications*, Holden-Day, San Francisco (1968).
[2]   R. B. Blackman and J. W. Tukey, *The Measurement of Power Spectra*, Dover (1958).
[3]   H. Akaike, On the use of an index of bias in the estimation of power spectra, *Ann. Inst. Statist. Math.*, Vol. 20 (1968) 55–69.
[4]   H. Akaike and Y. Yamanouchi, On the statistical estimation of frequency response function, *Ann. Inst. Statist. Math.*, Vol. 14 (1962) 23–56.

[5]   H. Akaike, Statistical measuremet of frequency response function, *Supplement III, Ann. Inst. Statist. Math.* (1964) 5–17.

[6]   H. Akaike, On the statistical estimation of the frequency response function of a system having multiple input, *Ann. Inst. Statist. Math.*, Vol. 17 (1965) 185–210.

[7]   H. Akaike, Statistical predictor identification, *Ann. Inst. Statist. Math.*, Vol. 22 (1970) 203–217.

[8]   H. Akaike, On a semi-automatic power spectrum estimation procedure, *Proc. 3rd Hawaii International Conference on System Sciences* (1970) 974–977.

[9]   H. Akaike, Autoregressive model fitting for control, *Ann. Inst. Statist. Math.*, Vol. 23 (1971) 163–180.

[10]   H. Akaike, On the use of a linear model for the identification of feedback systems, *Ann. Inst. Statist. Math.*, Vol. 20 (1968) 425–439.

[11]   H. Akaike, Some problems in the application of the cross-spectral method, *Spectral Analysis of Time Series* (B. Harris ed.), John Wiley (1967) 81–107.

[12]   H. Akaike, A method of statistical identification of discrete time parameter linear systems, *Ann. Inst. Statist. Math.*, Vol. 21 (1969) 225–242.

[13]   H. Akaike, On a decision procedure for system identification, *Preprints, IFAC Kyoto Symposium on System Engineering Approach to Computer Control* (1970) 485–490.

[14]   W. S. Widnall, *Applications of Optimal Control Theory to Computer Controller Design*, MIT Press, Cambridge, Massachusetts (1968).

[15]   R. E. Kalman, New development in systems theory relevant to biology, *Systems Theory and Biology* (M. D. Mesarović ed.), Springer-Verlag, New York (1968) 222–232.

Chapter 4

A SUCCESSFUL APPLICATION

In this chapter, the practical use of the statistical methods discussed in the previous chapter is explained by using a real example. This example is concerned with the realization of the statistical analysis and control of the cement kiln process and has already been reported by the present authors in a paper [1] appearing in Automatica, the journal of IFAC (International Federation of Automatic Control). Here we avoid technical details that are specific to the cement kiln process and keep discussions to the general conceptual level. This will make discussions accessible to more readers who are interested in the analysis and control of general statistical systems. In reading this chapter, it will be very effective if the reader returns to Chapter 3 whenever it becomes necessary. As in the preceding chapter, Programs 5.2.1 etc. denote computer programs given in Chapter 5.

4.1 Spectral Analysis of the Cement Kiln

Our research started by looking carefully at the actual behavior of a rotary kiln. Fig. 4.1-1 shows an example of the record of a measurement of a kiln obtained at the early stage of our research. The reader should refer, as required, to the explanation given in Chapter 2 of the definition and measuring point of each variable dealt with in this chapter. Fig. 4.1-2 shows a record of the same variable as that of Fig. 4.1-1 under ideal operating conditions. As is obvious from the comparison of these two figures, the state of the kiln shown by Fig. 4.1-1 is definitely out of control.

To understand the nature of this disturbed behavior more quantitatively, we calculated power spectra for some measurements of the kiln. These are shown in Fig. 4.1-3. The figure shows that every variable has large power in the very low frequency range near the direct current component with frequency zero, considerable power near the frequency of one cycle every two

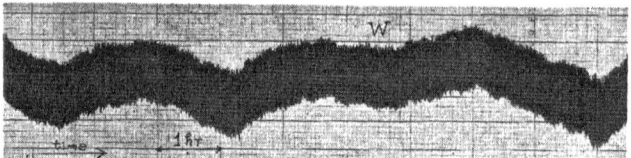

Fig. 4.1-1   An example of the record of the load power (W) of the kiln drive motor
under unstable operating condition.

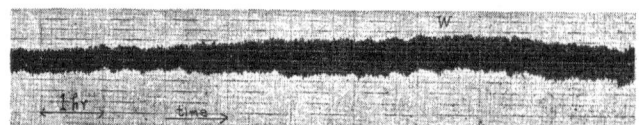

Fig. 4.1-2   An example of the record of the load power (W) of the kiln drive motor
under stable operating condition.

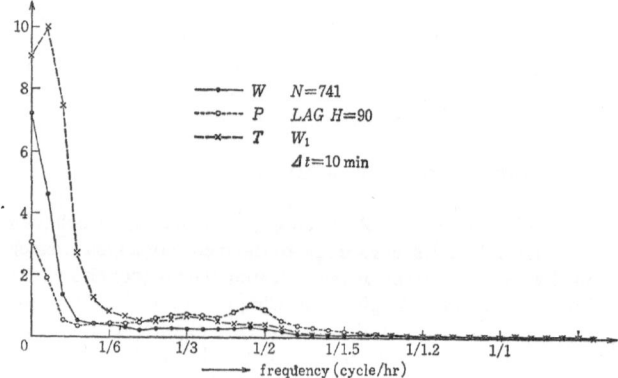

Fig. 4.1-3   Power spectra of controlled variables.

or three hours and looses power rapidly as the frequency is increased further. From this result, it is clear that our attention should be concentrated on the frequency range between direct current and one cycle per hour.

As the data used in this analysis has been sampled every 10 minutes the upper limit of the calculated spectrum reaches a frequency of one cycle every 20 minutes. The aliasing caused by the sampling must have some effect on the higher frequency side but its influence on the frequency range below one cycle per hour is expected to be small.

The fluctuation of the under cooler pressure (P) contains fairly significant higher frequency components. As will be shown by later analysis, the under cooler pressure responds sensitively to the control of the cooler grate speed by the operator, with a delay of a few minutes, and the final control system eventually adopted the sampling interval of 4 minutes.

In the early stage of analysis it was considered that practically satisfactory results were obtained by using data taken with a sampling interval of 10 minutes. This was because very low frequency fluctuation was dominant as was already seen by Fig. 4.1-3.

We used as our data only those that showed nearly stationary movements and were observed continuously for over 3 days. The spectrum was obtained by the method of 3.1.4 (Programs 5.1.1 and 5.2.1), and the maximum lag of the correlations calculated was 90. Estimates of power spectra can now also be obtained by Programs 5.3.1 and 5.4.1.

In the same way that the histogram show the frequency distribution of observed data in ordinary statistical analysis, the power spectrum shows the distribution of the total power of a time series over the frequency axis. In ordinary statistical analysis the covariance or the coefficient of correlation is used to analyze the linear relation between two variables, but in the present situation we use the cross-spectrum or coherency, instead. This means that we intend to investigate the linear relation between two stationary time series by looking at the relation between the corresponding components at each frequency. By using the result of this analysis we wish to find out what kind of relation holds between which pair of variables and, if possible, learn whether there is a possibility of suppressing the fluctuation in question by manipulating some variables properly. Fig.'s 4.1-4 and 4.1-5 show an example of the calculation of coherencies, corresponding to the squares of the correlation coefficients between each pair of frequency components, obtained by the method of 3.1.4 using Program 5.2.2.

It is clear from this result that in the frequency range in which we are interested, high coherencies are observed between variables representing the behavior of the kiln itself, such as the kiln end gas temperature (T), under cooler pressure (P), and the kiln drive power (W). This suggests that the

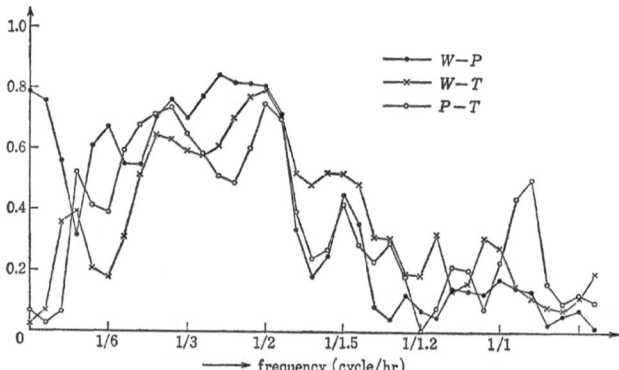

Fig. 4.1-4   Coherencies between controlled variables.

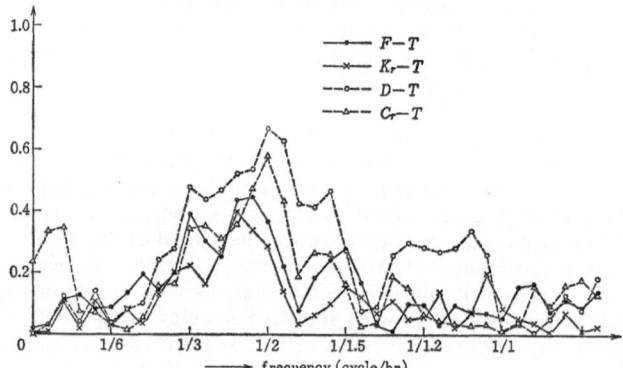

Fig. 4.1-5   Coherencies between the kiln end gas temperature (T) and the manipulated
variables.

variables are showing fairly similar movements in this frequency range, except for the effect of some time lags. It is now clear that each variable has a definite relation with the others, rather than moving independently.

In contrast to this, the coherencies between the manipulated variables, representing the action of the operator, and the above cited controlled variables are not very high, as can be seen from Fig. 4.1-5. This figure shows the coherencies between the manipulated variables and T.

Similar results were obtained for other controlled variables, such as W, P, etc. This indicates that a simple linear relation cannot be found between the manipulation by the operator and the movement of the variables that represent the state of the kiln. Does this mean that the manipulated variables are ineffective? Or does this show the variability of the manipulation by the operator? In fact, cross-spectral analysis does not provide any definite information to answer these questions. The multiple coherency between a controlled variable and the manipulated input variables shows a higher value than the simple coherency between a controlled variable and one of the manipulated variables; refer to 3.1.4 and Program 5.2.4. However, even this multiple coherency analysis cannot provide any answer to the previous questions. This difficulty is caused by the existence of the feedback between the variables. Before proceeding to the solution of this difficulty, let us see by a concrete example what is meant by feedback and why it causes the analytical difficulty. Incidentally we note here that by the result of 3.2.2 the computation of coherencies can now be done by using Programs 5.3.2 and 5.4.2.

A typical simultaneous movement of the kiln end gas temperature (T), under cooler pressure (P) and kiln drive power (W) is shown in Fig. 4.1-6. One possible interpretation of this figure is given by the following reasoning. Starting from the left end of Fig. 4.1-6, we first assume that, for some reason, the kiln end gas temperature (T) has risen above its normal level. This rise causes the rise of the kiln drive power (W) with a time delay of about 45 minutes. This interpretation is natural if we consider that the increase of W represents the increase of the activity of the kiln caused by the rise of the temperature.

Now the rise of W causes the rise of under cooler pressure (P), again with a delay of about 45 minutes. This interpretation is also natural if we think that temporarily a large mass of clinker is fed into the cooler when the kiln activity is increased. Thus we get a complete explanation of the downward flow of the fluctuation of the kiln condition accompanied with the downward flow of the mass inside the kiln.

On the other hand, Fig. 4.1-6 also suggests the possibility of another explanation of the fluctuation of the kiln condition due to the upward transmission of the fluctuation caused at the lower end of the kiln. First, the

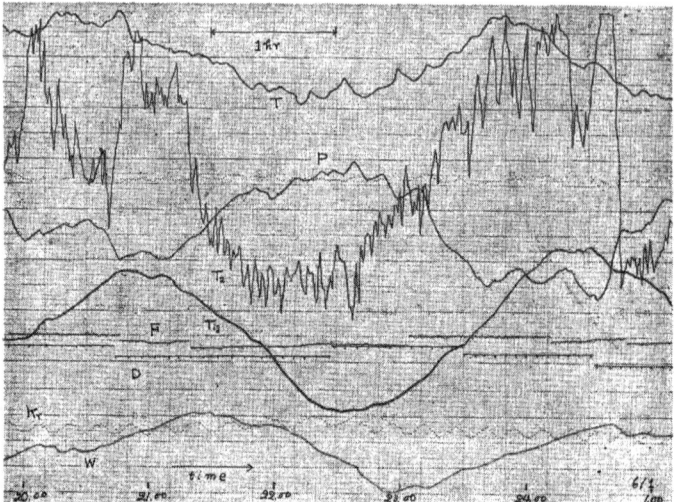

Fig. 4.1-6   A typical example of the record of variations of controlled and manipulated variables.

rise of the under cooler pressure P causes the lowering of kiln end gas
temperature T. If we consider the rise of P as an indicator of the increase of the
resistance of the flow of secondary air, the air blown into the kiln through the
cooler grate, the above interpretation looks natural as the rise of P causes poor
burning of fuel and at the same time causes a decrease of heat transfer from the
clinker to the secondary air, thus causing lowering of T. Also it is possible to
think that the rise of W causes the lowering of T after some delay. It may also
be possible to consider that the rise of P leads to the fall of W. It is rather
difficult to confirm the validity of this last interpretation at this moment.

   These possible upward flows complete the link of the upward and
downward flows by producing effects with signs opposite to those assumed by
the initial state of the original downward flow and provide an explanation of
how the fluctuation of the kiln condition having a period of about 3 hours is
formed naturally.

   Our problem here is the determination of the cause of all the fluctuations.
This, however, cannot be done easily. Any fluctuation of T, P or W can either

be the cause or the consequence of fluctuations of other variables. Therefore the validity of the above type of interpretation cannot be confirmed by looking only at the pattern of the data. We are faced here with an object whose behavior is impossible to interpret definitely, no matter how much time we spend in visually studying the data.

This problem comes from the difficulty of applying ordinary cross-spectral analysis to this type of problem. For example, to see the influence of the fluctuation of P on the behavior of W, consider the single input single output system with P as the input and W as the output. By applying the procedure of 3.1.4 (Program 5.2.3) to the estimation of the frequency response function of this system one gets only an incomprehensible result.

An example of the estimate of the impulse response function was obtained by computing the inverse Fourier transform of an estimate of the frequency response function obtained by this procedure and a part of the result is illustrated in Fig. 4.1-7. This result shows a significant response on the negative side of the time axis, which shows that the assumed uni-directional system is not physically realizable. This means that we need a more sophisticated statistical method for the analysis of a system having feedback.

4.2 Selection of Variables

A real kiln is an enormous object, the full length being rather more than a hundred meters, and the physical and chemical processes taking place inside it are, as was discussed in Chapter 2, quite complicated. Therefore, when the control of a kiln is contemplated, the decision on what variables to measure becomes a serious problem. The maninpulated variables which are those which will actually be manipulated in reality can be identified clearly but it is not easy to define the controlled variables that will properly represent the state of the kiln. However, despite this problem, human operators have been

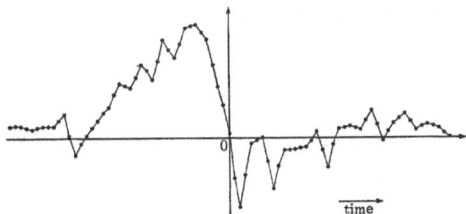

Fig. 4.1-7   An example of unsuccessful estimation of an impulse response function.

controlling the kiln, by using some measurements and observing the internal state of the kiln through the observation window. From this it is expected that, when a control is implemented, only a finite number of observation points will suffice, that is, it is not necessary to treat the kiln as a continuous body or a distributed system. However, in our case, with the development of measurement techniques, the number of kiln measurement points was once increased to 34. Did we use all the information from so many points? Actually the measurements to be used in control were limited to the following:

1) Variables chosen based on the consideration of the physical and chemical characteristics of the kiln; for example, the above cited T.

2) Variables that are known empirically to correspond closely to various conditions of the kiln; for example, W.

3) Those with low observation noise. From this point of view the burning zone temperature and the kiln end gas oxygen content have been excluded temporarily.

4) Those that no other measurement that shows similar movement is retained. For example, under certain measurement conditions, the secondary air temperature shows fluctuations similar to those of the under cooler pressure, except for the opposite sign. Moreover, it also shows significantly high frequency fluctuations. Thus the secondary air temperature was tentatively excluded. The intermediate gas temperature which shows the temperature at the lower end of the calcining zone was also excluded from consideration as it also showed a fairly strong linear relation with the kiln end gas temperature and was apt to be influenced by the burning zone radiation.

These judgements were made gradually by putting together previous experience, direct observations of the data, and results of the analysis of correlations among the variables, etc. In particular, 3) and 4) are essential when we try to identify a model by the statistical method with the observational data. By the way, as was discussed already in 3.5.5, FPEC can be used for the selection of variables. For an example of this use of FPEC, the reader is directed to reference [9] of Chapter 3. Thus we decided to carry on our investigation by limiting the variables to the following set.

First, as the controlled variables we adopted

kiln drive power	(W)	(kW)
under cooler pressure	(P)	(mm H_2O)
kiln end gas temperature	(T)	(°C)

and as the manipulated variables

fuel rate	(F)	(m^3·hr)
kiln rotation speed	(K_r)	(r.p.h.)
kiln end gas damper opening	(D)	(°)

cooler grate speed (C_r) (r.p.m.) .

In the following discussion, we sometimes use the symbol inside the first parentheses to represent the corresponding variable. The symbol inside the second parentheses is the measurement unit of each variable. Common sense suggests that among the above mentioned manipulated variables both the fuel rate (F) and the kiln end gas damper opening (D) will act directly on the kiln end gas temperature (T), while the cooler grate speed (C_r) will act directly on the under cooler pressure (P). However, the effect of the change of the kiln rotation speed (K_r) is extremely complicated and its prediction by simple guessing is impractical. This point was already briefly discussed in Chapter 2. Obviously an instantaneous increase in K_r leads directly to an increase in W. However, this is merely a simple electrical response and cannot be considered as the response of W indicative of the physical and chemical state of the kiln. Whatever reasoning we use, an analysis based simply on common sense cannot reveal the existence of a manipulated variable that can be used to control W directly.

The idea of using the kiln drive power (W) as an item of information for the control of the kiln process was first introduced by one of the present authors [2]. The importance of this variable will become clear through the following discussion. As can be seen by Fig. 4.1-1, the actual kiln drive power shows a cyclic fluctuation with a roughly constant amplitude, due to the rotation of the kiln having a frequency of less than one cycle per minute. In the following analysis, the effect of this cyclical fluctuation is removed by smoothing the values of the kiln drive power sampled with a smaller sampling internal by an appropriate numerical filter to define W to be used in the analysis.

4.3   Analysis of the Kiln as a Feedback System

In this section, we discuss the result obtained by applying the method of analysis of the feedback system, discussed in 3.3.2, to the kiln data. Although various methods have been tried during the actual process of our research, here we will give, in Table 4.3-1, a part of the result obtained by applying the method of fitting an autoregressive model to the kiln data. This method was discussed in the latter half of 3.3.2, following equation (16). (See Program 5.3.3.) In this case, the length of the data was 741 and the order M of the autoregression was 5.

The (i, j)-th element of Table 4.3-1 shows the value of $q_{ij}(f)$ of (22) of 3.3.2 divided by $p_{ii}(f)$ of (21). (It is printed out under the title of differential relative power contribution in the output of Program 5.5.3.) Therefore, this value shows the ratio of the power of the fluctuation of the i-th variable caused by

Table 4.3-1   An example of relative power contribution analysis (at frequency 1 cycle per 2 hours).

F= 15

DIFFERENTIAL RELATIVE POWER CONTRIBUTION

MATRIX   7 X 7

		W 1	P 2	T 3	F 4	Kr 5	D 6	Cr 7
W	1	0.35059D 00	0.20662D 00	0.28029D 00	0.21030D-01	0.23354D-01	0.35233D-01	0.82884D-01
P	2	0.93997D-01	0.51903D 00	0.27611D 00	0.34592D-01	0.63419D-03	0.29846D-01	0.45794D-01
T	3	0.62720D-01	0.77543D-01	0.71108D 00	0.29527D-01	0.13251D-01	0.66717D-01	0.39161D-01
F	4	0.20699D-01	0.73173D-02	0.73934D-01	0.82147D 00	0.45340D-01	0.22485D-01	0.87595D-02
Kr	5	0.19305D-01	0.76011D-01	0.59368D-01	0.45332D-02	0.78547D 00	0.59787D-02	0.49335D-01
D	6	0.51405D-01	0.73579D-01	0.18545D 00	0.15712D-01	0.48226D-02	0.62862D 00	0.40616D-01
Cr	7	0.57719D-01	0.11464D 00	0.12395D 00	0.27565D-01	0.11761D-02	0.54278D-01	0.62067D 00

the fluctuation, or noise, inherent in the j-th variable to the total power of the i-th variable at frequency f. Now, the variables are defined as $x_1 = W$, $x_2 = P$, $x_3 = T$, $x_4 = F$, $x_5 = K_r$, $x_6 = D$, and $x_7 = C_r$.

Actually the cooler grate drive current was used as x_7 in place of the cooler grate speed (C_r). Several problems arise because of this, but we will not discuss them here. Instead, we will continue our discussion by pretending $x_7 = C_r$.

Table 4.3-1 shows the result corresponding to the 15-th frequency among the 90 frequencies that divide the whole frequency band into 90 equal intervals. As the present sampling interval is 10 minutes,

$$f = \frac{15}{90} \times \frac{1}{2 \times 10} = \frac{1}{120} \quad \text{(cycle/minute)}$$

and hence Table 4.3-1 gives the result at the frequency of one cycle per two hours.

In Table 4.3-1, a diagonal element close to 1 indicates that the fluctuation of the corresponding variable at that frequency is due almost entirely to the noise originating in the variable itself. One remarkable point is that the diagonal elements corresponding to the manipulated variables x_4, x_5, x_6 and x_7, have large values. This suggests that, as far as the linear model is concerned, the fluctuation in the manipulated variables caused by the intervention of human operator is not responding to fluctuations originating in the kiln process. Some exceptions are the response of $x_6 = D$ to $x_3 = T$ and the response of $x_7 = C_r$ to $x_2 = P$ and $x_3 = T$. On the contrary, some controlled variables are responding to fluctuations originating in other variables. For example, $x_1 = W$ is influenced mainly by the fluctuations originating in $x_2 = P$

and $x_3 = T$. We can also see that $x_2 = P$ is influenced by $x_3 = T$ and $x_1 = W$. However, most of the fluctuation of $x_3 = T$ is due to the noise inherent in itself. We can see some influences of $x_7 = C_r$ on $x_1 = W$ and that of $x_6 = D$ on $x_3 = T$, while other influences from manipulated variables on controlled variables are very small. This result suggests, for example, the possibility of decreasing the fluctuation of $x_1 = W$ by controlling $x_2 = P$ and $x_3 = T$ properly and also the possibility of producing a good effect on $x_2 = P$ by appropriate control of $x_1 = W$ and $x_3 = T$, etc. At least it is empirically obvious that $x_3 = T$ (kiln end gas temperature) rises when $x_4 = F$ (fuel rate) is increased. Thus the ineffectiveness of the manipulation by the operator, especially of $x_4 = F$, as shown in Table 4.3-1, is rather striking.

If these observations are valid, there is a possibility of realizing a control superior to that exerted by a human operator by using an appropriately designed automatic control. However, the variables $x_1 = P$ and $x_3 = T$ should not be handled separately. The control must be designed by taking these variables as a set of mutually interacting variables. The main emphasis of the control should first be placed on the reduction of the fluctuation of $x_3 = T$ that seems to be relatively unaffected by other variables, while producing influences on other controlled variables. For this, the manipulation of $x_4 = F$ and $x_6 = D$ is expected to be useful. To decrease the fluctuations of $x_1 = W$ which can be considered as the final indicator of the state of the process, we have to control the influence of the fluctuation of $x_2 = P$. The variable $x_7 = C_r$ will be useful for this purpose.

We decided first to identify the response of $x_3 = T$ to $x_4 = F$. Fig. 4.3-1 shows the response of T to F obtained by using the equation (26) of 3.3.2. (See Program 5.3.4.) It can be seen that the response of T to F is of integral type. This observation clarifies the effectiveness of introducing a differential type

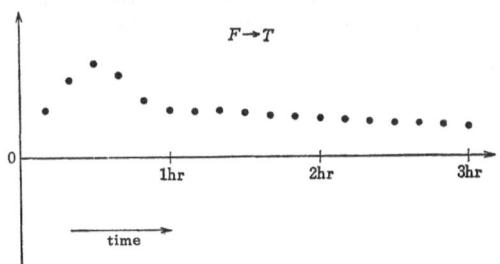

Fig. 4.3-1   The response of the kiln end gas temperature (T) to the fuel rate (F).

negative feedback between T and F which increases or decreases the fuel rate in proportion to the negative of the time difference of the kiln end gas temperature (T), where the time differences is taken to be the change of T over 10 minutes.

A more detailed analysis of the data showed that feedback of this type would not cause instability in the range between the zero frequency and a frequency of about one cycle per hour, even when the gain of the feedback was very high. To confirm the validity of this conclusion, we implemented an experimental fuel control of this type in a real kiln. In this experiment, the gain of the feedback loop was increased up to the point where the fluctuation of F attained its empirically allowable maximum amplitude. In spite of the increase of the gain, the kiln behavior was extremely stable, and we could not observe any significant oscillation of T. Obviously this control system could never be effective in the low frequency range near the zero frequency, as information about any low frequency oscillation was attenuated by the process of taking the difference of T. However, this was the first experimental confirmation of the result obtained by our statistical analysis and suggested the possibility of realizing a control based on the result of such an analysis.

The defect in the design of the above control of T was that it could not control certain type of fluctuations of the kiln which were clearly visible in the record of the kiln drive power (W). We introduced an empirically designed feedback between W and F, but this also did not produce a satisfactory result. Our main problem was then how to realize the control of W.

The physical significance of W is not so clear as that of T. The kiln rotation speed K_r is generally controlled to maintain a constant value, except for possible minute fluctuations. The fluctuation of W in this circumstance can, as we discussed before, be considered as an indicator of the kiln activity. When the burning zone grows significantly, W is expected to rise. Further, the change of the mass distribution, caused, for example, by dropping of the kiln coating, will obviously influence W directly. Thus, controlling W is equivalent to controlling nearly all the condition in the kiln. Since in the preceding experiment, a relatively fast change in W that could not be handled by the control of T by F was observed, we were forced to find a means to produce an influence on W more directly. Does there really exist a measure for the control of W which is a quantity that depends on the whole of the mass distribution inside the gigantic kiln? The result of our statistical analysis explained with Table 4.3-1 already provides an answer to this question.

Recall the observation that W was influenced significantly by the fluctuation of P, as well as that of T. That is, if we control the under cooler pressure (P) appropriately, there is a possibility of realizing a control of W that will cover a considerable portion of the fluctuation of W that could not be handled through the control of T. However, is it really possible to influence the complicated behavior of W through controlling P? By the result of our analysis, the response of P to W corresponding to the response of F to T discussed above, was similar to that shown in Fig. 4.3-2. This type of response was observed consistently in a number of examples. It suggests that a rise of P leads to a relatively speedly fall of W. If this observation is valid, this is a major discovery. The truth of this result could not be confirmed by simple logical analysis. Accordingly we talked with the most skilled kiln operator to get some clues about the validity of this observation.

Generally a kiln operator's testimony contains a wealth of experience, so that listening to the whole story makes one feel that controlling the kiln by using a computer would definitely be impossible. However, among the many comments, this statement was made;

"When P starts to rise, W starts to fall. Leaving a kiln in this state will eventually lead to a lot of problems. Thus we increase the cooler rotation speed (C_r) to counteract the rise of P".

This almost confirmed both the validity of the result of our analysis and the possibility of using the analysis to realize control system.

When the analysis was started, the amount of cement output was above the specification of the kiln. In particular, the clinker cooler was running at its maximum speed under a full load and it was practically impossible to implement control of the cooler. If the present observation of the effect of P on W is correct, this shows that one of the main causes of the difficulty in operating the kiln was the cooler system. The common practice up to that time

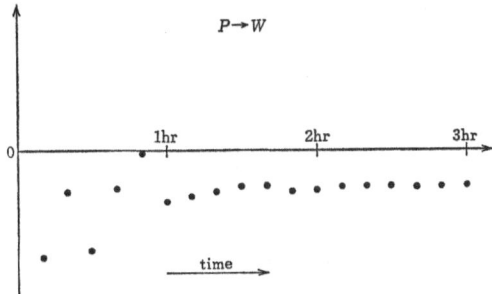

Fig. 4.3-2   The response of the kiln drive power (W) to the under cooler grate pressure (P).

was to treat the control of the cooler and that of the kiln separately. Our result was the first to show objectively the necessity of controlling the kiln through the manipulation of the cooler.

The necessity of improving the cooler was thus recognized and the remodeling required for the realization of the kiln and cooler combined control was performed. At this point an IBM 1800 computer was added to the computer system to prepare for the realization of a new overall control.

From the result corresponding to that of Table 4.3-1 in the very low frequency domain it was observed that the kiln end gas damper opening (D) was producing effects on the movement of the controlled variables. An example is shown in Table 4.3-2. Although the result is partly due to the fact that the actual manipulation of the damper was done only intermittently we may consider that the result reflects degree of contribution of D to the determination of the state of the kiln in the direct current range. Also the fuel rate was found to produce a similar effect on the kiln. These results indicate that there is a possibility of improving the behavior of the kiln at the very low frequency range by manipulating the variables D and F properly.

Here we show in Table 4.3-3 the matrix of the coefficients of correlation obtained by normalizing the variance-covariance matrix of the innovations or the white noise vector of the fitted autoregressive model. This is to see the appropriateness of the assumption of uncorrelatedness between the components, the basic condition for the validity of the above analysis. It is clear that the correlations are very low. The length N of the data used for this calculation was 741.

Before finishing our discussion of the result of our analysis, we must note that in the real application, the validity of each result was repeatedly checked with data of similar observations.

Table 4.3-2   An example of relative power contribution analysis ($f=0$).

F=   0

DIFFERENTIAL RELATIVE POWER CONTRIBUTION						
MATRIX	7 W	X P	T	F	Fr	D	Cr
	1	2	3	4	5	6	7
W 1	0.19218D 00	0.49822D-01	0.47601D-01	0.16661D 00	0.20023D-01	0.52367D 00	0.95236D-04
P 2	0.16580D 00	0.13003D-02	0.12708D-01	0.22989D 00	0.33103D-01	0.54131D 00	0.15897D-01
T 3	0.25097D 00	0.67683D-01	0.29789D 00	0.18150D 00	0.11593D-01	0.14557D 00	0.44797D-01
F 4	0.23038D-01	0.41322D-01	0.34872D-02	0.24471D 00	0.11471D 00	0.57246D 00	0.27501D-03
Fr 5	0.65978D-01	0.61432D-01	0.27395D-03	0.63812D-01	0.17015D 00	0.63767D 00	0.68712D-03
D 6	0.56729D-01	0.53074D-01	0.22982D-01	0.27184D-01	0.56390D-01	0.80196D 00	0.23612D-02
Cr 7	0.16628D 00	0.10802D 00	0.85919D-01	0.72950D-03	0.10937D-01	0.16877D 00	0.45935D 00

Table 4.3-3   An example of correlation matrix of innovations.

NORMALIZED SD

MATRIX	7 X 7 W 1	P 2	T 3	F 4	Kr 5	D 6	Cr 7
W · 1	0.100000 01	-0.703640-01	-0.189270 00	0.523120-01	0.379990 00	0.242650 00	-0.637010-01
P 2	-0.703640-01	0.100000 01	0.489150-02	0.100430 00	0.938860-01	-0.820130-02	0.251060-01
T 3	-0.189270 00	0.489150-02	0.100000 01	-0.371220-01	-0.547460-01	-0.229170 00	-0.164930-01
F 4	0.523120-01	0.100430 00	-0.371220-01	0.100000 01	0.974630-01	0.199310 00	0.661060-01
Kr 5	0.379990 00	0.938860-01	-0.547460-01	0.974630-01	0.100000 01	0.355990 00	-0.281370-01
D 6	0.242650 00	-0.820130-02	-0.229170 00	0.199310 00	0.355990 00	0.100000 01	-0.181510-01
Cr 7	-0.637010-01	0.251060-01	-0.164930-01	0.661060-01	-0.281370-01	-0.181510-01	0.100000 01

4.4 Realization of the Control System

Finally we come to explain our experience of the application of the statistical controller design discussed in 3.5 to the realization of the kiln control system. At the beginning of this whole investigation, we had the following apprehensions.

1. An operator manipulates the kiln on the basis of extremely complicated judgements. Wouldn't it be a short cut to success to make an effort to realize a logical circuit by implementing a computer program that approximates the working of this complicated judgemental system of an operator?

2. The behavior of the kiln is extremely complicated and delicate. For example, it becomes impossible to continue operation without proper control when the kiln is exposed to a rain shower. Is it really possible to realize effective control by using a linear model for this kind of situation?

Our experience in handling statistical problems suggested the following answers to these questions.

1. We cannot develop a drastic countermeasure against a complicated statistical phenomenon only by accumulating empirical measures based on the understanding of the phenomenon, without looking at it on the basis of some particular principle. Therefore, the use of empirical logic should be kept to the minimum and one should not think the use of it at the beginning of the investigation.

2. As long as there exists a linear relation that is observable by eye, such as the relation between the increase of the fuel rate (F) and the rise of the kiln end gas temperature (T), there is a sufficient probability of implementing an appropriate feedback even when the noise is very strong. However, the effectiveness of a model identified by the statistical

method can only be proved through the actual implementation of the control.

Thus, we first spent much of our effort in the analysis of the kiln, as we already discussed in the preceding sections up to 4.3. As a result we were able to develop a basic understanding of the coupling between the kiln and cooler behavior and eventually obtained a simple and lucid image of the kiln that clearly superceded the conventional understanding represented by the so-called "complicated judgements" of the operator. The only thing left now was to prove the effectiveness of this image by realizing the control system.

In the real application, the time sampled control system was adopted. This system observes and controls a continuous time system at constant time intervals and the manipulated variables are kept constant during the time interval. The performance of this type of system is discussed, for example, in reference [14] of Chapter 3. Here we will simply assume that the sampling rate is high enough for the frequency range of interest to make the discrete time model a reasonable approximation to the sampled control system.

4.4.1   Implementation of local control of the cooler

The remodeling of the kiln was performed on the basis of the result of the analysis discussed in the preceding sections up to 4.3 and the introduction of an IBM 1800 computer prepared the conditions for the realization of the control. However, to check the appropriateness of the result of the statistical analysis, we first tried a simple local stabilizing control of the cooler. This control system was a simplification of the complete kiln-cooler control system and we will not discuss it in detail here. It is a system based on the model obtained from the record of the kiln behavior under the control of a human operator and the feedback loop was realized by inserting a computer between the under cooler pressure (P) and the cooler rotation speed (C_r) to reduce the fluctuation of P. The control was performed every two minutes. The effect was greater than expected and the operator confirmed that the operation of the kiln became very easy. In this case, all the manipulated variables except C_r were under the operator's control.

The significance of the effect of this cooler control could also be judged from the fact that this experiment caused the delay of the implementation of the next step of the overall control. Operators simply did not want to disconnect the experimental cooler control. Partial confirmation of the result of our statistical analysis was thus almost complete.

All the experiments of local controls discussed thus far were concerned with the control of fluctuations at rather high frequencies. However, the main problem was the control of fluctuation in the low frequency range. As significant changes in conditions often occur during the periods when the operator is working the kiln, it is very hard to get a long stationary record of observations for the determination of the model of the kiln process. This suggests that we cannot expect to estimate our model very accurately in the very low frequency range. However, as we already saw in the power spectrum of Fig. 4.1-3, the fluctuation of the kiln shows significant power at low frequencies near the zero frequency. Will the statistically determined model actually realize an effective control of the fluctuation in this frequency range? As the time of the application of the overall control drew near we continued our preparation with a prayer for success.

4.4.2 Control of the kiln-cooler system

It was widely known that, for the design of a multivariate control system, the empirical approach based on the classical frequency response function or the transfer function would be impractical. The method based on the state space representation of the system was supposed to replace this classical approach. The new control theory was developed on the basis of this state space representation and its realization was supported by the introduction of computers. A research group in IBM was trying to put this method into practice. Among the output of this group, the report on the application of the state space design method to the cement process by K. Y. Wong and others [3] produced a strong stimulus to our work. However, there were significant differences between the work of our group and that of the IBM group in the following 3 points.

1) The example of application of the state space design method to the cement process by Dr. Wong *et al.* did not handle the cooler and the kiln as one system. This seems to be a natural consequence of the omission of the stage of the analysis of the kiln. The necessity of combining the kiln and the cooler into one system was the most important conclusion reached by our present statistical analysis.

2) In Wong *et al.*'s report, the determination of the model was performed by using the differenced series of observations. This is incompatible with our attitude that places decisive importance on the low frequency range behavior.

3) In the above report the determination of the order of the model was carried out subjectively and was a source of serious difficulty in practice. For this problem, we have an almost perfectly objective method discussed in 3.5.5 and therefore there is no difficulty in practical application.

Obviously there are many other differences between the results reported by the two groups concerning the problems and their solutions in the implementation of control. However, here we limited our attention to the basic aspects described above.

In January of 1970, a good review of computer control in the cement industry was published by V. A. Kaiser [4]. According to this review, complete automation of the cement process was considered to be impossible at that time, because of the difficulties of obtaining measurements and determining the model. As was already mentioned, our controlled variables were the three variables, W, P, and T, and the manipulated variables were the four variables, F, K_r, D, and C_r. The time interval for observation and control was four minutes. The determination of the model was done almost automatically following the procedure of 3.5.5. About two days' records of the kiln were used as the original data, where all the manipulated variables were under the control of human operators and nearly stationary operations were maintained. The lengths of data thus obtained were about 720. The orders of the models determined by equation (4) of 3.5.5 for data of this length were usually 5 or 6. Of course this is a value determined not only by the length of data N but by the characteristics of the kiln and noise sources. The equation (19) of 3.5.2 which is the basis of the controller design is determined by equation (3) of 3.5.5.

The controller was designed by following the procedure defined by the formulas (23) to (28) of 3.5.3. Q_1 and R that define the quadratic criterion of the design were determined according to the method discussed in the latter half of 3.5.4. We chose 10 as the length I of the control span for dynamic programming by equation (27) of 3.5.3. Actually the value 20 was considered to be desirable for I of our case. However, because of the limitation of the numerical accuracy of the computation using the IBM 1800, we reduced it to 10.

The gain of the kiln rotation speed controller was reduced to 0 by making the diagonal element of the matrix R corresponding to the kiln rotation speed (K_r) very large. This was because there were various difficulties related to the manipulation of the kiln rotation speed.

Therefore, the first control was realized with the kiln rotation speed (K_r) kept constant. The results of the simulation study of the kiln behavior showed that there was a possibility for the variance and the mean square of the fluctuation of each controlled variable to fall below one fifth of its original value. An example of the result of the gain calculation using the IBM 1800 is shown in Table 4.4.2-1. Fig.'s 4.4.2-1 and 4.4.2-2 provide an example of comparison of the responses of the system with and without control, where an impulsive noise was added to the model. In this example, the variable W_2, to be explained later, is used in place of W.

Table 4.4.2-2 shows the data of the first experiment of the kiln and cooler combined control. A part of the corresponding record of operations is shown

Table 4.4.2-1 A sample output of the gain calculation.

```
                    ***  KILN 1 A,B AND G CHECK  ***

        LL=  4  NR=  3  MT=  6

         A(M,I,J)     FILE NO. = 1001            B(M,I,J)     FILE NO. = 1011
             W            P            T              F            K            D            C
M=  1    W   0.824370E 00 -0.219830E-02 -0.457220E-01   W  0.103830E-01  0.457660E-01  0.129690E 00 -0.261820E-01
         P  -0.958810E-01  0.140530E 01 -0.242510E 00   P  0.147260E 01  0.281540E 00 -0.241080E 00 -0.100960E 01
         T   0.105500E-01  0.185090E-02  0.135740E 01   T -0.394510E-01  0.229410E-01 -0.155580E 00 -0.518630E-02
M=  2    W            P            T                        F            K            D            C
         W  -0.433540E 00 -0.858820E-02 -0.162980E 00   W -0.370840E 00  0.229980E 00 -0.323160E 00  0.240220E-02
         P  -0.915820E-01 -0.575290E 00 -0.725310E-01   P  0.178230E 00 -0.245750E 00  0.483760E 00  0.336160E 00
         T  -0.122130E-01 -0.345670E-02 -0.526150E 00   T -0.231270E-01 -0.889460E-02 -0.198790E-01  0.201110E-02
M=  3    W            P            T                        F            K            D            C
         W   0.821690E 00 -0.443510E-02  0.115110E 00   W  0.693920E 00  0.455640E 00  0.284270E 00 -0.222170E-01
         P   0.282680E 00  0.201000E 00  0.60577CE 00   P -0.574380E 00  0.358870E 00 -0.210670E 00 -0.338350E 00
         T  -0.100150E-02  0.300360E-02  0.354310E 00   T  0.627190E 00 -0.804130E-01  0.231690E 00 -0.220000E-01
M=  4    W            P            T                        F            K            D            C
         W   0.336750E-01 -0.215230E-02 -0.818370E-01   W  0.797720E 00 -0.336640E 00 -0.167680E 00  0.320980E-01
         P  -0.130180E 00 -0.792320E-01 -0.867090E 00   P  0.286850E 00 -0.355950E 00  0.483980E 00  0.794680E 00
         T   0.597860E-02 -0.142340E-01 -0.176290E 00   T -0.372030E-01 -0.418400E-02  0.255680E 00  0.560690E-01
M=  5    W            P            T                        F            K            D            C
         W   0.201250E 00  0.900980E-02  0.224610E-01   W -0.815210E-01  0.578330E-01  0.236370E 00 -0.202940E-01
         P   0.320890E 00  0.347740E-01  0.201540E 00   P -0.202570E 01  0.977820E-01 -0.656170E 00  0.120550E 00
         T  -0.127860E-02  0.384270E-02  0.407060E-01   T -0.309450E 00  0.603510E-01 -0.262020E 00  0.203310E-02
M=  6    W            P            T                        F            K            D            C
         W  -0.150920E 00 -0.864700E-02  0.173520E 00   W -0.253310E 00 -0.399790E 00 -0.268640E 00 -0.317720E-02
         P  -0.200050E 00 -0.368110E-01  0.134980E 00   P  0.391160E 00 -0.200520E 00 -0.168790E 00 -0.317180E-01
         T  -0.516770E-02  0.107800E-01 -0.688900E-01   T -0.638510E-02 -0.286860E-01 -0.142470E-01 -0.141710E-01

         G(M,I,J)     FILE NO. = 1021
             W            P            T                        W            P            T
M=  1    F  -0.797560E-02  0.842277E-03 -0.122989E-01   F  -0.156522E-01  0.285896E-01 -0.152680E 01
         K  -0.428817E-03  0.662520E-05  0.106823E-02   K  -0.119737E-02  0.847000E-05  0.593260E-03
         D   0.100580E-01 -0.177052E-01 -0.151127E 00   D   0.285785E-01 -0.245954E-01 -0.238005E 00
         C  -0.605563E-01  0.344785E 00 -0.156880E 00   C  -0.685360E-01  0.786551E-01 -0.443212E-01
M=  2    F  -0.233072E-02  0.670030E-03 -0.123503E-01   F  -0.168037E-01  0.970841E-04 -0.126413E-01
         K  -0.568011E-03 -0.312936E-04  0.852275E-03   K  -0.205168E-03  0.433790E-05  0.462837E-03
         D   0.801776E-01 -0.199741E-01 -0.156238E 00   D   0.103025E-01 -0.190514E-01 -0.220442E 00
         C   0.202556E-02  0.310515E 00 -0.109690E 00   C  -0.112161E 00 -0.792960E-02 -0.756510E-01
M=  3    F  -0.310765E-02  0.482640E-03 -0.135376E-01   F  -0.348355E-02  0.106082E-05 -0.921545E-02
         K  -0.134022E-02 -0.314314E-04  0.603340E-03   K   0.156235E-02 -0.293978E-05  0.508037E-03
         D   0.131806E 00 -0.252308E-01 -0.103887E 00   D   0.430358E-01 -0.507333E-02 -0.136157E 00
         C  -0.791788E-02  0.188004E 00 -0.303069E-01   C  -0.273685E-01 -0.180363E-01 -0.600213E-01

   // END
```

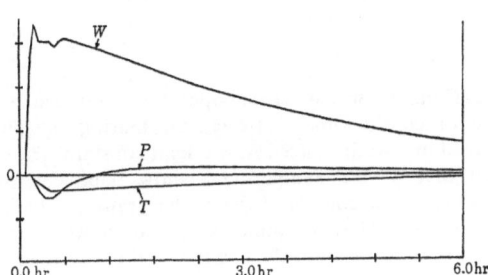

Fig. 4.4.2-1 Response of the non-controlled system (simulation).

$\xi(n) = \Phi\xi(n-1)$

$\xi_1(n) = W$ (kiln drive power: kW)

$\xi_2(n) = P$ (under cooler pressure: mmH$_2$O)

$\xi_3(n) = T$ (kiln end gas temperature: °C)

Minimum scale ranges $W: -1 \sim +1$ $P: -5 \sim +5$ $T: -1 \sim +1$

An impulsive disturbance of size +5 (kW) was added to W.

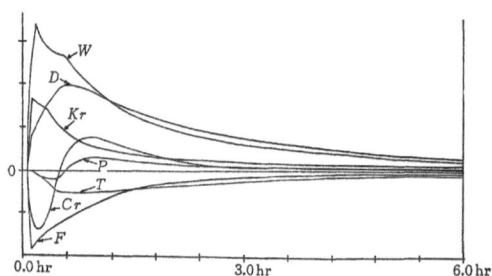

Fig. 4.4.2-2   Response of the controlled system (simulation).
$$\xi(n)=\varPhi\xi(n-1)+\varGamma y(n-1)$$
$$y(n-1)=G\xi(n-1)$$

$\xi_1(n)=W$ (kiln drive power: kW)
$\xi_2(n)=P$ (under cooler pressure: mmH$_2$O)
$\xi_3(n)=T$ (kiln end gas temperature: °C)
$y_1(n)=F$ (fuel rate: m^3/hr)
$y_2(n)=K_r$ (kiln rotation speed: r.p.m.)
$y_3(n)=D$ (kiln end gas damper opening: °)
$y_4(n)=C_r$ (cooler gate speed: r.p.m.)

    Minimum scale ranges W: $-1\sim+1$ P: $-5\sim+5$ T: $-1\sim+1$ F: $-0.1\sim+0.1$
K_r: $-0.1\sim+0.1$ D: $-0.1\sim+0.1$ C_r: $-0.1\sim+0.1$

    An impulsive disturbance of size $+5$ (kW) was added to W.

in Fig. 4.4.2-3. From this we can see that operation continued safely in spite of the existence of considerable disturbances. During the period of this experiment, the kiln rotation speed K_r was kept constant. The control of the fuel (F) and that of the damper (D) show relatively low frequency movements, whereas the control of the cooler grate speed (C_r) shows high frequency movements. These movements agree quite well with the expectations obtained from the result of the analysis of the preceding Section 4.3.

In this case, the determination of the model and the design of the control system was done using data from kiln No. 2. However, because of conditions at the control site, the experiment was carried out with the kiln No. 1 which had exactly the same size and structure as kiln No. 2. In this first experiment, the control system based on our simple model safely operated the complicated enormous kiln amid considerable disturbances for a few days, until it was halted because of mechanical problems. It became clear from this experiment that the design procedure had considerable safety margin, as the resulting control was rather insensitive to the difference between the kiln No. 1 and No.

Table 4.4.2-2 An example of the kiln-cooler control record.

date	water content (%)	fuel consumption (kl)	output (ton)	fuel consumption per output (l/ton)	remarks
7.6	35.3	159.8	1,200	133.2	12.00 on-line start
7	35.1	159.1	1,213	131.2	on-line
8	35.1	160.1	1,222	131.0	on-line
9	35.1	160.7	1,229	130.8	on-line
10	35.0	160.5	1,242	129.1	on-line
11	—	—	—	—	9.40 on-line off**
15	34.8	156.5	1,237	126.6	manual
16	35.0	158.1	1,282	128.2	manual
17	34.2	156.7	1,263	124.0	15.00 on-line start
18	34.5	155.1	1,252	123.9	on-line
19	34.5	156.5	1,252	125.0	on-line
20	34.4	156.9	1,262	124.0	on-line
21	34.3	156.8	1,256	124.8	on-line
22	33.8	155.3	1,274	121.9	on-line
23	—	—	—	—	13.00 on-line off***

*Kiln rotation speed was kept constant at about 60 rotations per hour.
**Stopped because of trouble in the raw material feeder.
***Stopped because of an accident to the cooler blower motor.

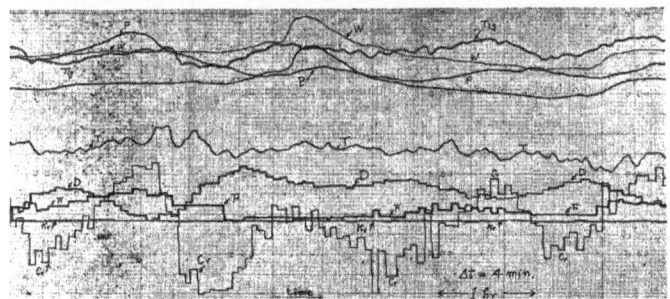

Fig. 4.4.2-3   A part of the operation record corresponding to Table 4.4.2-2.

2. Another point that was made clear by this experiment was the fact that this control system could operate the kiln successfully for a sufficiently long period, if only mechanical problems as shown in Table 4.4.2-2 did not occur. A comparison of control by a human operator and our control based on a record obtained at the time of this experiment is shown in Fig. 4.4.2-4. The differences of the power spectra of controlled variables demonstrates the effectiveness of the computer control in eliminating the low frequency fluctuations. The spectra of the human control shown in Fig. 4.4.2-4 were obtained from data of control by a human operator which were used for the determination of the model.

The success of this experimental computer control produced various items of information useful for the improvement of the kiln system. This point will be discussed later. The experience obtained by this experiment also suggested the possibility of developing a control system with higher practical utility by adding appropriate control of the kiln rotation speed (K_r). The first problem in implementing the control of K_r is the existence of the electrical coupling between K_r and W.

As was discussed before, from the point of view of constructing a process model, this coupling means the introduction of a noise into W as an indicator

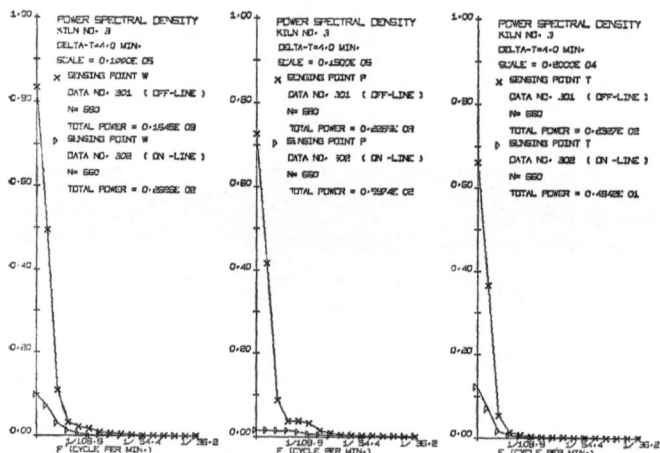

Fig. 4.4.2-4   Comparison between human and computer control by power spectra. ✕: man, ▷: computer.

of the physical state of the process. Accordingly we decided to eliminate this influence tentatively by performing a calculation that will cancel the momentary change of W caused by the increase or decrease of the electrical load when a discontinuous change occurs in K_r. The variable thus derived from W was denoted by W_2. In the case of the example shown in Table 4.4.2-3, the design and the implementation of the control was done using W_2 in place of W. The human control data shown in the Table is based on the data used to determine the model for computer control. Using this data, the variance of each controlled variable, such as W, P, T etc., was reduced by computer control to about one fifth of its magnitude under human control. This result corresponds well to the results of the preceding simulation. In the example of Table 4.4.2-2, the kiln was completely under computer control, whereas for the result given in Table 4.4.2-3 the operator was allowed to manipulate the kiln rotation speed K_r when necessary. In the case of Table 4.4.2-3, the small value of the variance of K_r under computer control indicates that K_r was scarcely manipulated.

Thus our expectation was realized almost completely. Obviously this was only a success in principle and to put the control completely into practice several further refinements are required. This point will be discussed in more detail later. Nevertheless it would be fair to say that the correctness of our attitude taken toward the questions stated at the beginning of this Section 4.4 has been proved by this realization.

Table 4.4.2-3   Comparison between human and the computer control (variances).

	human control data length $N=680$**		computer control data length $N=660$**		
	mean	variance	mean	variance	set point
W (kW)	180.010	164.519	204.662	26.265	——
W_2 (kW)	168.188	50.848	204.795	29.204	203.260
P (mmH$_2$O)	110.188	228.596	122.345	59.746	124.100
T (°C)	172.398	23.277	173.612	4.843	173.740
F (m^3/h)	7.024	0.015	7.686	0.0045	7.653
K_r (r.p.h.)	55.244	2.301	57.429	0.147	58.405
D (°)	45.319	4.181	52.140	0.262	50.728
C_r (r.p.m.)	21.911	3.034	22.586	8.370	23.798

*Data used for the determination of model.
**$\Delta t = 4$ min.

4.5 Information Obtained by the Success of the Experiment

The success of the experimental computer control of the kiln has opened up almost unlimited possibilities for its future development. This is due to the possibility of obtaining useful information through observation of the kiln during its operation without intervention by the human operator. This was made possible only by the introduction of complete computer control. As was expected already in the stage of the analysis of the kiln behavior the operator was actually generating very high noise levels that significantly reduced the signal to noise ratios (S/N) of the measurements of the kiln. The following are examples that explain this point in concrete terms.

1)   It was observed clearly that a change of about 5% in the voltage of the electrical power supply affected the kiln condition significantly. To properly compensate for the influence of the variation of the voltage of this order it was necessary to control K_r beyond the level of precision of the available control devices of the kiln rotation speed. Under human control, the low freuqncy fluctuation of each record was so large that it was impossible to recognize the effect of this voltage fluctuation and the necessity of this kind of control was completely ignored.

2)   It was confirmed that one of the essential causes of the fluctuation of W was the irregular movement of the electric dust collector that collects the dust floating inside the kiln and sends it back into the kiln as part of the raw material (the secondary material). Hitherto, the operator took necessary corrective action for each particular situation and it was impossible to distinguish the effect caused by the irregular movement of the electric dust collector from those caused by other noise sources.

Both of the above problems 1) and 2) were left unnoticed for many years under operation by a human operator. The above observations lead the management quickly to a decision to introduce necessary improvements to the related devices and their instrumentations.

The above results all demonstrate the effectiveness of the statistically identified model of the kiln. When the model is adequate, it is expected, as was discussed at the end of 3.5.3, that the one-step prediction error by the model will approximately be a white noise. When a disturbance occurs that changes the characteristics of the system, the effect will easily be detected by the deviation of the sequence of the one-step prediction errors from whiteness. It is not easy to detect a change in the movements of the controlled variables that represent the behavior of the kiln itself, as the response of the kiln is generally very slow, i.e. with a large time constant, and at the same time the normal range of its fluctuations is very wide. However, once the fluctuation is transformed into a sequence of one-step prediction errors, it behaves like a white noise. Thus it can be expected that the detection of the gradual change caused by the change of the characteristics of the system will be made possible by a simple averaging operation. This expectation has been validated empirically in many situations.

Fig. 4.5-1 shows an example of the record of one-step prediction errors before and after the numerical low-pass filtering which suppresses high frequency components. The right half of Fig. 4.5-1 shows the data of the transition phase where the kiln is drifting away from normal operating conditions. The high frequency fluctuation of each record is significantly reduced by low-pass filtering and we can clearly see the change in the characteristics of the kiln behavior.

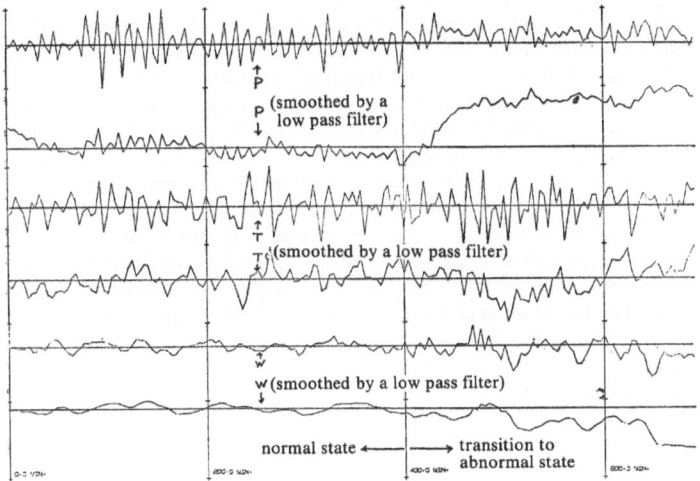

Fig. 4.5-1   One-step prediction errors and their smoothing. (Transition from a normal to an abnormal condition.)

This result suggests the possibility of using the information supplied by the prediction error sequence for the implementation of a higher order control that includes the adjustment of the control to compensate for the change in the characteristics of the kiln, that is, the so-called dual control. An example of this type of control will be discussed shortly. Here we will discuss another example to show the effectiveness of the present modeling approach for the detection of the deviation of a system from its normal operating condition.

In one experiment, the behavior of the kiln showed significant deviation from the behavior expected from the model. The pattern of the deviation shown by the behavior of the prediction error sequences suggested the shortage of oxygen within the air supply. By investigation, it became clear that the cooler damper that controlled the air flow within the cooler was fixed in an extremely abnormal position. By correcting the damper position, the kiln quickly returned to its normal state.

As this example shows a clearly defined appropriate model can be extremely effective for the detection of disorder within the system. A human operator often executes control by performing the necessary corrective adjustments and unknowingly continues operation of the system which is already in an abnormal condition. This is a commonly observed phenomenon, not restricted to the kiln control, that often provides a justification for the presently adopted control by forming an obstacle to the proper recognition of its limitation. In such a situation, an objective confirmation of the structure of the system under observation, such as the one realized by the fitting of our present model, often leads to a breakthrough in conventional control that has been previously developed mainly on an empirical basis.

The success of our experiment allowed us to determine the direction of the necessary structural improvement of the kiln system and the elimination of external noise sources. If these actions were carried out properly, the control system thus far developed would become practically applicable. However, there are also other kinds of unavoidable noises that are spontaneous in character. In addition to noises caused by accidents, there is a sudden change in the mass distribution inside the kiln due to the dropping of a slab of cement coating of the kiln wall. This causes changes in the operating condition of the kiln that often produce effects that cannot be properly compensated by linear control based on the present time invariant model.

The only action available to cope with this situation has been to reduce the kiln rotation speed K_r and wait for the kiln to return to its normal state of operation. The manipulation of K_r produces an extremely complex influence on the kiln. Statistical analysis using the linear model did not reveal any significant influence of K_r on the kiln behavior. However, in the real operation of the kiln, it is well-known that the manipulation of K_r plays a decisive role.

The manipulation of K_r is mainly concerned with the control of the kiln when it is in a non-stationary state and the response of the kiln to the change of K_r is usually highly non-linear. Thus we cannot realize a proper control of K_r using a linear model. Also it must be remembered that it is impractical to adjust K_r continuously.

To overcome these difficulties we considered the use of the information provided by the sequence of one-step ahead prediction errors discussed earlier. We noticed a fast response of the kiln drive power (W) to the change of mass distribution inside the kiln and decided to increase or decrease K_r slightly only when a proper transformation of the sequence of prediction errors showed an abnormal deviation having an extremely low probability to be observed under normal conditions. The direction of the increase and decrease was empirically chosen to pull the kiln back to the normal state and, after the kiln returned to the normal state, K_r was returned to its original set point. In practical applications W was replaced by W_2 which was obtained by eliminating the influence of the electric coupling between W and K_r. The robustness of the control system has been improved remarkably by the addition of this K_r control. However, this K_r control will obviously affect the performance of the original linear control system. To compensate for this, for example, some adjustment of the fuel rate (F) that responds to the manipulation of K_r was added. The introduction of such minor adjustments allowed longer operation of the kiln without human intervention with only a slight deterioration of the quality of control under normal operating conditions.

An example of the analysis of the operation record under this final control system is shown in Fig. 4.5-2. Table 4.5-1 shows the standard deviations of the variables. Fig. 4.5-3 is the schematic diagram of the present control system. The control computer IBM 1800 executes the computations discussed in this book. Compared with the old control system shown in Fig. 2.3.3-3 the new system is quite simple in its structure.

With the implementation of the K_r control, our control system is almost ready for practical applications. However, some serious problems still remain. These are the choice of the set point, the reference value such that the controller responds to a deviation of the variable from this value, for each variable and the implementation of the automatic start up control that brings the kiln into normal operating condition from its resting state [5]. These problems become significant when we try to put the control system into practice. The development of statistical methods for the solution of these problems is a subject of future study. At present, all we can do is to develop some practically useful procedures empirically.

These problems, the above discussed K_r control and the use of W_2 as the

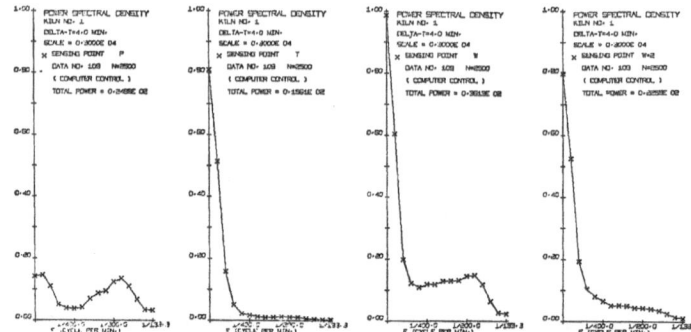

Fig. 4.5-2 Power spectra for an operation record taken over an extended period of
time.

Table 4.5.1 Statistics for the long term operation (means and standard deviations).

	first week* data length $N=2500$		second week** data length $N=2500$		third week* data length $N=2400$	
	mean	standard deviation	mean	standard deviation	mean	standard deviation
W (kW)	215.867	6.013	215.526	7.715	215.705	9.497
W_2 (kW)	216.399	4.750	215.975	6.327	215.594	9.425
P (mmH$_2$O)	98.616	4.990	99.096	6.629	105.141	9.243
T (°C)	144.218	3.952	140.919	3.241	145.547	2.671
F (m^3/h)	7.194	0.102	7.172	0.121	7.215	0.108
K_r (r.p.h.)	59.961	0.397	59.572	0.503	58.342	0.626
D (°)	59.884	1.379	58.292	1.763	57.284	0.641
Cr (r.p.m.)	28.336	1.822	28.060	1.529	28.348	2.329

*The type of cement was changed from normal cement to quick-setting cement during the
on-line control.
**A reverse change to the above took place.
Above data for three weeks were selected from the available set of data so as to obtain records
of almost equal length.

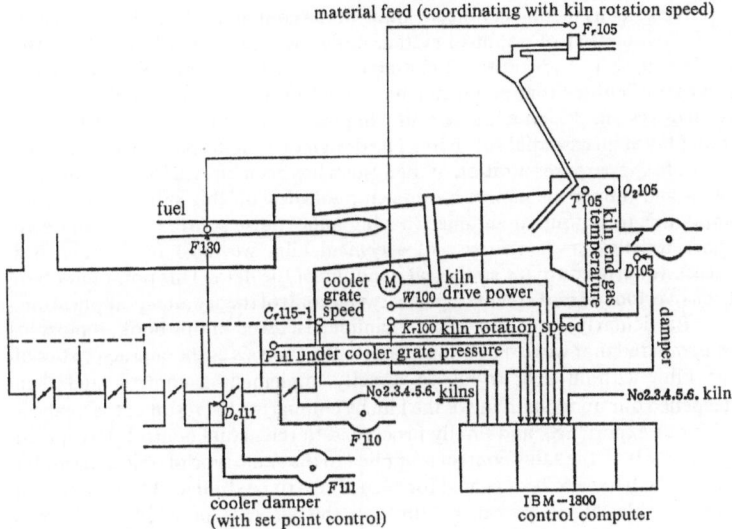

material feed (coordinating with kiln rotation speed)

Fig. 4.5-3   Schematic diagram of the kiln cooler control system.

basis for control, show that in the study of real objects there are always aspects which are beyond the limit of the direct application of already existing general theory. Even some of the statistical methods discussed in this book have been newly developed during the process of this study. We hope that our experience will not only show the effectiveness of the proper application of already existing theory but also explain that real objects are always full of challenging new problems.

4.6   Concluding Remarks

Before concluding we note the following, though it sounds rather obvious.

In using statistically estimated values, the effect of statistical errors must always be taken into account. For this purpose, it is advisable to check variability and its effect on the design of a control system by using data of similar observations or of simulations. One must avoid putting too much faith in one single result without careful checking. A final judgement can only be based on the performance of the result when it is put into practical use.

Now, let us reconsider the meaning of the content of the present chapter. It is obvious that our control system design was consistently based on the confirmation of the statistical characteristics of the system. This approach solved difficulties that could not be treated by the conventional empirical control system design explained in Chapter 2. The statistical processing of data played an essential role here. The design of a modern control system by the state space representation of the system has been known for more than 10 years and numerous papers have been published on this subject. When we apply this technique to an object having many noise sources with unknown characteristics, as in the case of a cement kiln, we must first establish a successful procedure for statistical handling of the data. This point has often been overlooked and created a gap between control theory and its application.

Particularly in the case of the example discussed in this book, it must be remembered that we first developed a detailed analysis of the characteristics of the kiln, without ever trying to directly implement a control, and then deepened our understanding of the kiln by comparing the result of the analysis with our experience, and finally proceeded to the actual control. Except for the case where the same control is applied to the same type of object, a similar process will have to be repeated for each object to be studied. The selection of the variables itself is realized through this process of analysis. As was discussed in Chapter 1, the success of the initial analysis or control leads to further analysis and control of noise sources and thus the control is extended to its limit.

Some important aspects of the present study that have not been touched before are as follows:

1)   the establishment of close cooperation between the groups of people concerned with management, instrumentation and control, and statistical methodology, and

2)   the advancement of the technical level and the development of human resources within the related groups realized through the process of this cooperative research.

The first aspect 1) was the basis of the success of the present study, and 2) was a priceless by-product. These aspects will be recognized in almost all similar researches and developments. We sincerely hope that similar studies will be conducted in many fields and, with the increasing use of the computer, lead to the development of the theory and application of the statistical analysis and the control of dynamic systems.

It is most desirable that this book is used in an environment where aspect 1) is assured. But we also hope that this book itself will serve as a catalyst to create such an environment.

REFERENCES

[1] T. Otomo, T. Nakagawa and H. Akaike, Statistical approach to computer control of cement rotary kilns, *Automatica*, Vol. 8, No. 1 (1972) 35–48.

[2] T. Nakagawa, A study on the Control of Rotary Kiln, Doctoral Dissertation, University of Tokyo (1964). (in Japanese)

[3] K. Y. Wong, K. M. Wiig and E. J. Albritton, Computer control of the Clarksville cement plant by state space design method, A paper presented at IEEE Cement Industry Technical Conference, St. Louis, U.S.A., May 20–24 (1968).

[4] V. A. Kaiser, Computer control in the cement industry, *Proc. IEEE*, Vol 58 (1970) 70–77.

[5] Y. Yagihara and T. Kominami, On the computer control of cement rotary kilns. A paper presented at the 28th Symposium on Cement Production Technique, November 1971. (in Japanese)

Chapter 5

COMPUTER PROGRAMS (TIMSAC PACKAGE)

The practical utility of a statistical procedure can be appreciated only by applying it to real data. Therefore, for the ease of application of the procedure described in this book, the necessary computer programs written in FORTRAN are provided in this chapter. These programs are aimed at the explanation of the principles for the construction of the necessary computational procedures. It is hoped that programs suitable for each particular combination of the object of application and computer will be developed based on these programs. Naturally, programs developed only for the purpose of the kiln control are not included in the present package.

The programs are written in HITAC 8300/8400/8500 FORTRAN IV. It has been checked that these programs can run on ordinary IBM computers without any modification. Each program is structured in such a way that basic tasks are described in the form of subroutines and the main program plays the role of the flow chart that connects these tasks. The use of each program is explained by the comments given in the program.

To save space, subroutines used in more than one programs are collected in 5.6.

Expressions such as IMPLICIT COMPLEX * $16(X-Z)$ found in programs indicate that variables with names beginning with an alphabetic character between X and Z are complex variables defined with 16 bytes (1 byte=8 bits). If COMPLEX is replaced by REAL, the variables are real variables. Needless to say, for a computer system which cannot directly handle complex variables, the programs must be rewritten using only real variables.

The programs given in this chapter have been developed particularly for the convenience of readers of this book, and thus they are independent of the system of programs applied to the actual control of the cement kiln discussed in Chapter 4. Accordingly, there might be some incomplete points in the programs. We hope that readers would develop proper modifications when

131

necessary.

The programs given in this chapter take the form of a program package for the analysis and control of time series. Thus, for the convenience of future reference, we name it the TIMSAC (Time Series Analysis and Control) Package.

5.1 Covariance Function Computation

The calculation of the estimates (hereafter "estimates" will be omitted, unless there is a danger of confusion) of the autocovariance function or the cross-covariance function from a given set of data forms the basis of the procedure developed in this book. There are two ways to calculate the covariance function. The first method (direct method) calculates covariances directly according to the definition and the second method (FFT method) uses the FFT (Fast Fourier Transform). By this latter method the Fourier transforms for given data are obtained by FFT. For a pair of variables we replace the Fourier transform of one of the variables by its conjugate complex and form the product of this with the other. The inverse Fourier transform of this frequency domain function gives the desired covariance function between the two variables.

The merits and demerits of the two methods are as follows:

1) The merit of the direct method is the simplicity of the program. The demerit is the computing time required when the length of the data is large.

2) The merit and demerit of the FFT method is just the reverse of the above. Thus the method is computationally advantageous when the data length is large.

[Programs]

5.1.1 AUTCOR Autocovariance Function Computation (uni-variate; direct method)

5.1.2 MULCOR Cross-covariance Function Computation (multivariate; direct method)

5.1.3 FFTCOR Cross-covariance Function Computation (bi-variate; FFT method)

Program 5.1.3 can be used as a subroutine for other programs. The main part of this program is based on a mimeographed note on the ALGOL procedure for the FFT calculation which was made available by Dr. Gordon Sande, then a graduate student and now at Statistics Canada, when one of the authors was visiting Princeton University, U.S.A., in 1966–7. The reader is referred to [1] for a general reference on FFT.

```
      PROGRAM AUTCOR
C     PROGRAM 5.1.1    AUTO CORRELATION
C     THIS PROGRAM REQUIRES FOLLOWING INPUTS:
C     N: LENGTH OF DATA
C     LAGH: MAXIMUM LAG
C     DFORM: INPUT FORMAT SPECIFICATION STATEMENT IN ONE CARD,
C     FOR EXAMPLE
C     (8F10.4)
C     (X(I),I=1,N): ORIGINAL DATA.
C     THE OUTPUTS ARE AUTOCOVARIANCES (CXX(I); I=0,LAGH) AND.
C     AUTO CORRELATIONS (NORMALIZED COVARIANCES).
      DOUBLE PRECISION CXX,CN,CX0
      DIMENSION X(2000),CXX(501),CN(501)
      DIMENSION DFORM(20)
C     INITIAL CONDITION INPUT AND PRINT OUT
      READ(5,1) N,LAGH
      WRITE(6,50)
      WRITE(6,51)
      WRITE(6,52) N,LAGH
C     INPUT FORMAT SPECIFICATION
      READ(5,4) (DFORM(I),I=1,20)
    4 FORMAT(20A4)
C     ORIGINAL DATA INPUT AND PRINT OUT
      READ(5,DFORM) (X(I),I=1,N)
      WRITE(6,53)
      WRITE(6,54)
      DO 220 I=1,N
  220 WRITE(6,55) I,X(I)
C     MEAN DELETION
      CALL SMEADL(X,N,XMEAN)
C     AUTO COVARIANCE COMPUTATION
      LAGH1=LAGH+1
      CALL CROSCO(X,X,N,CXX,LAGH1)
C     NORMALIZATION
      CX0=CXX(1)
      CALL CORNOM(CXX,CN,LAGH1,CX0,CX0)
C     AUTO COVARIANCE PRINT OUT
      WRITE(6,162) N,LAGH,XMEAN
      WRITE(6,163)
      CALL PRCOL2(CXX,CN,1,LAGH1,1)
C     AUTO COVARIANCE PUNCH OUT
      WRITE(7,1) N,LAGH
      WRITE(7,2) (CXX(I),I=1,LAGH1)
      STOP
    1 FORMAT(10I5)
    2 FORMAT(4D20.10)
   50 FORMAT(1H1,13HPROGRAM 5.1.1,3X,16HAUTO CORRELATION)
   51 FORMAT(1H0,17HINITIAL CONDITION)
   52 FORMAT(1H0,2HN=,I5,5X,5HLAGH=,I5)
   53 FORMAT(1H0,13HORIGINAL DATA)
   54 FORMAT(1H0,4X,1HI,6X,4HX(I))
   55 FORMAT(1H ,I5,2X,F10.4)
  162 FORMAT(//1H0,14HAUTOCOVARIANCE,5X,6HCXX(I),5X,2HN=,I5,5X,5HLAGH=,I
     A5,5X,5HMEAN=,E15.5)
  163 FORMAT(1H0,4X,1HI,5X,6HCXX(I),8X,10HNORMALIZED)
      END
```

```
      PROGRAM MULCOR
C     PROGRAM 5.1.2   MULTIPLE CORRELATION
C     THIS PROGRAM REQUIRES FOLLOWING INPUTS:
C     N: LENGTH OF DATA
C     K: DIMENSION OF THE OBSERVATION VECTOR
C     LAGH: MAXIMUM LAG
C     ISW: ISW=1...ROWWISE DATA INPUT
C          ISW=2...COLUMNWISE DATA INPUT
C     DFORM: INPUT FORMAT SPECIFICATION STATEMENT IN ONE CARD,
C     FOR EXAMPLE
C     (8F10.4)
C     (X1(S,I); S=1,...,N, I=1,...,K): ORIGINAL DATA MATRIX.
C     THE OUTPUTS ARE (CIJ(L): L=0,1,...,LAGH) (I=1,...,K; J=1,...,K),
C     WHERE CIJ(L)=COVARIANCE(XI(S+L),XJ(S)),
C     AND THEIR NORMALIZED (CORRELATION) VALUES.
      DOUBLE PRECISION C1,C2,CN1,CN2,C0,CX0,CY0
      DIMENSION X1(2000,10)
      DIMENSION X(2000),Y(2000)
      DIMENSION C1(501),C2(501),CN1(501),CN2(501)
      DIMENSION SM(10),C0(10)
      DIMENSION DFORM(20)
C     INITIAL CONDITION INPUT AND OUTPUT
      READ(5,1) N,LAGH,K,ISW
      LAGH1=LAGH+1
      WRITE(6,50)
      WRITE(6,51)
      WRITE(6,52) N,LAGH,K
C     INITIAL CONDITION PUNCH OUT
      WRITE(7,1) N,LAGH,K
C     INPUT FORMAT SPECIFICATION
      READ(5,4) (DFORM(I),I=1,20)
    4 FORMAT(20A4)
C     ORIGINAL DATA INPUT AND OUTPUT
      GO TO(8,9),ISW
    8 DO 208 I=1,N
  208 READ(5,DFORM) (X1(I,II),II=1,K)
      GO TO 400
    9 DO 209 II=1,K
  209 READ(5,DFORM) (X1(I,II),I=1,N)
  400 WRITE(6,53)
      WRITE(6,54)
      WRITE(6,154)
      DO 220 I=1,N
  220 WRITE(6,55) I,(X1(I,II),II=1,K)
C     MEAN DELETION
      DO 300 II=1,K
      DO 310 I=1,N
  310 X(I)=X1(I,II)
      CALL SMEADL(X,N,XMEAN)
      DO 320 I=1,N
  320 X1(I,II)=X(I)
      SM(II)=XMEAN
  300 CONTINUE
C     COVARIANCE COMPUTATION
      DO 10 II=1,K
      DO 110 I=1,N
  110 X(I)=X1(I,II)
```

```
C     AUTO COVARIANCE COMPUTATION
      CALL CROSCO(X,X,N,C1,LAGH1)
C     NORMALIZATION
      C0(II)=C1(1)
      CX0=C0(II)
      CALL CORNOM(C1,CN1,LAGH1,CX0,CX0)
C     AUTO COVARIANCE PRINT OUT
      WRITE(6,162) II,II,SM(II)
      WRITE(6,163)
      CALL PRCOL2(C1,CN1,1,LAGH1,1)
C     AUTO COVARIANCE PUNCH OUT
      WRITE(7,1) II,II
      WRITE(7,2) (C1(I),I=1,LAGH1)
      IF(II.EQ.1) GO TO 10
      IM1=II-1
      DO 11 JJ=1,IM1
      DO 120 I=1,N
  120 Y(I)=X1(I,JJ)
C     CROSS COVARIANCE COMPUTATION
      CALL CROSCO(X,Y,N,C1,LAGH1)
      CALL CROSCO(Y,X,N,C2,LAGH1)
C     NORMALIZATION
      CX0=C0(II)
      CY0=C0(JJ)
      CALL CORNOM(C1,CN1,LAGH1,CX0,CY0)
      CALL CORNOM(C2,CN2,LAGH1,CX0,CY0)
C     CROSS COVARIANCE PRINT OUT
      WRITE(6,165) II,JJ
      WRITE(6,166)
      CALL PRCOL4(C1,CN1,C2,CN2,1,LAGH1,1)
C     CROSS COVARIANCE PUNCH OUT
      WRITE(7,1) II,JJ
      WRITE(7,2) (C1(I),I=1,LAGH1)
      WRITE(7,1) JJ,II
      WRITE(7,2) (C2(I),I=1,LAGH1)
   11 CONTINUE
   10 CONTINUE
      STOP
    1 FORMAT(10I5)
    2 FORMAT(4D20.10)
   50 FORMAT(1H1,36HPROGRAM 5.1.2    MULTIPLE CORRELATION)
   51 FORMAT(1H0,17HINITIAL CONDITION)
   52 FORMAT(1H0,2HN=,I5,5X,5HLAGH=,I5,5X,2HK=,I5)
   53 FORMAT(1H0,13HORIGINAL DATA)
   54 FORMAT(1H0,4X,1HI,2X,8HX1(I,II))
  154 FORMAT(1H0,16X,91H1        2        3        4        5
     A  6        7        8        9       10)
   55 FORMAT(1H ,I5,2X,10F10.4)
  162 FORMAT(//1H0,14HAUTOCOVARIANCE,5X,6HCIJ(L),5X,2HI=,I5,5X,2HJ=,I5,5
     AX,5HMEAN=,E15.5)
  163 FORMAT(1H0,4X,1HL,5X,6HCIJ(L),8X,10HNORMALIZED)
  165 FORMAT(//1H0,16HCROSS COVARIANCE,5X,6HCIJ(L),5X,2HI=,I5,5X,2HJ=,I5
     A)
  166 FORMAT(1H0,4X,1HL,5X,6HCIJ(L),8X,10HNORMALIZED,4X,6HCJI(L),8X,10HN
     AORMALIZED)
      END
```

```
      PROGRAM FFTCOR
C     PROGRAM 5.1.3    AUTO AND/OR CROSS CORRELATIONS VIA FFT.
C     THIS PROGRAM COMPUTES AUTO AND/OR CROSS
C     COVARIANCES AND CORRELATIONS VIA FFT.
C     IT REQUIRES FOLLOWING INPUTS:
C     ISW: ISW=1...AUTO CORRELATION OF X (ONE-CHANNEL)
C          ISW=2...AUTO CORRELATIONS OF X AND Y (TWO-CHANNEL)
C          ISW=4...AUTO,CROSS CORRELATIONS OF X AND Y (TWO-CHANNEL)
C     LD: LENGTH OF DATA
C     LAGH: MAXIMUM LAG
C     DFORM: INPUT FORMAT SPECIFICATION STATEMENT IN ONE CARD,
C     FOR EXAMPLE
C     (8F10.4)
C     (X(I); I=1,LD): DATA OF CHANNEL X
C     (Y(I); I=1,LD): DATA OF CHANNEL Y (FOR ISW=2 OR 4 ONLY)
      IMPLICIT REAL*8(A-H,O-Y)
      IMPLICIT COMPLEX*16(Z)
      REAL*4 DFORM
      DIMENSION X(2048),Y(2048),Z(2048),ZS(1025)
      DIMENSION CN1(501),CN2(501)
      DIMENSION DFORM(20)
C     INITIAL CONDITION INPUT AND PUNCH OUT
      READ(5,1) ISW,LD,LAGH
      WRITE(6,50)
      WRITE(6,51)
      WRITE(6,52) ISW,LD,LAGH
      WRITE(7,1) LD,LAGH
      LAGH1=LAGH+1
      ND=LD+LAGH1
C     N2P, N: DEFINITION
      I0=1
   10 IR1=2**I0
      IF(IR1-ND) 11,12,12
   11 I0=I0+1
      GO TO 10
   12 N2P=I0
      N=2**N2P
      NP1=N+1
      NP2=N+2
      M=N/2
      M1=M+1
      AN=N
      ALD=LD
      ALD1=1.0/(AN*ALD)
C     INPUT FORMAT SPECIFICATION
      READ(5,4) (DFORM(I),I=1,20)
    4 FORMAT(20A4)
C     ORIGINAL DATA INPUT AND OUTPUT
      DO 20 I=1,N
      X(I)=0.0
   20 Y(I)=0.0
      READ(5,DFORM) (X(I),I=1,LD)
      IF(ISW.EQ.1) GO TO 200
      READ(5,DFORM) (Y(I),I=1,LD)
  200 WRITE(6,53)
      IF(ISW.NE.1) GO TO 201
      WRITE(6,54)
      CALL PRCOL1(X,1,LD,0)
      GO TO 202
  201 WRITE(6,55)
      CALL PRCOL2(X,Y,1,LD,0)
C     MEAN DELETION
  202 CALL DMEADL(X,LD,XMEAN)
      IF(ISW.EQ.1) GO TO 203
      CALL DMEADL(Y,LD,YMEAN)
```

```
C     DOUBLE PRECISION COMPLEX REPRESENTATION
203   DO 31 I=1,N
 31   Z(I)=DCMPLX(X(I),Y(I))
C     FOURIER TRANSFORM OF Z
      ISG=-1
      CALL MIXRAD(Z,N,N2P,ISG)
      IF(ISW.NE.1) GO TO 204
C     RAW SPECTRUM COMPUTATION
      DO 32 I=2,M
      X(I)=DREAL(Z(I))**2+DIMAG(Z(I))**2
      NI=NP2-I
 32   X(NI)=X(I)
      X(1)=DREAL(Z(1))**2
      X(M1)=DREAL(Z(M1))**2
      GO TO 205
C     DECOMPOSITION AND RAW SPECTRUM COMPUTATION
204   DO 125 I=2,M
      NI=NP2-I
      ZI=Z(I)
      ZNI=Z(NI)
      RF=DREAL(ZI)
      SF=DIMAG(ZI)
      RG=DREAL(ZNI)
      SG=DIMAG(ZNI)
      XI=RF+RG
      XNI=SF-SG
      Z(I)=DCMPLX(XI,XNI)
      X(I)=0.25*(XI**2+XNI**2)
      X(NI)=X(I)
      YI=SF+SG
      YNI=RF-RG
      Z(NI)=DCMPLX(YI,YNI)
      Y(I)=0.25*(YI**2+YNI**2)
      Y(NI)=Y(I)
125   CONTINUE
      X(1)=DREAL(Z(1))**2
      Y(1)=DIMAG(Z(1))**2
      X(M1)=DREAL(Z(M1))**2
      Y(M1)=DIMAG(Z(M1))**2
      IF(ISW.NE.4) GO TO 205
C     RAW CROSS SPECTRUM COMPUTATION
      DO 126 I=2,M
      NI=NP2-I
126   ZS(I)=0.25*Z(I)*Z(NI)
      ZS(1)=DREAL(Z(1))*DIMAG(Z(1))
      ZS(M1)=DREAL(Z(M1))*DIMAG(Z(M1))
C     AUTO COVARIANCE COMPUTATION
205   DO 33 I=1,N
 33   Z(I)=DCMPLX(X(I),Y(I))
C     FOURIER TRANSFORM
215   CALL MIXRAD(Z,N,N2P,ISG)
      II=1
      DO 34 I=1,LAGH1
 34   X(I)=DREAL(Z(I))*ALD1
      X0=X(1)
      AMEAN=XMEAN
C     NORMALIZATION
 36   CX0=X(1)
      CALL CORNOM(X,CN1,LAGH1,CX0,CX0)
C     AUTO COVARIANCE PRINT OUT
      WRITE(6,162) II,II,AMEAN
      WRITE(6,163)
      CALL PRCOL2(X,CN1,1,LAGH1,1)
```

```
C       AUTO COVARIANCE PUNCH OUT
        WRITE(7,1) II,II
        WRITE(7,2) (X(I),I=1,LAGH1)
        IF(ISW.EQ.1) GO TO 300
        IF(II.EQ.2) GO TO 216
        II=2
        DO 35 I=1,LAGH1
   35 X(I)=DIMAG(Z(I))*ALD1
        YO=X(1)
        AMEAN=YMEAN
        GO TO 36
  216 IF(ISW.NE.4) GO TO 300
C       CROSS COVARIANCE COMPUTATION
        DO 127 I=2,M
        NI=NP2-I
        Z(I)=ZS(I)
  127 Z(NI)=DCONJG(ZS(I))
        Z(1)=ZS(1)
        Z(M1)=ZS(M1)
C       FOURIER TRANSFORM
        CALL MIXRAD(Z,N,N2P,ISG)
        DO 41 I=1,LAGH
        I1=I+1
        J1=NP1-I
        X(I1)=DREAL(Z(I1))*ALD1
   41 Y(I1)=DREAL(Z(J1))*ALD1
        X(1)=DREAL(Z(1))*ALD1
        Y(1)=X(1)
C       NORMALIZATION
        CALL CORNOM(X,CN1,LAGH1,X0,Y0)
        CALL CORNOM(Y,CN2,LAGH1,X0,Y0)
C       CROSS COVARIANCE PRINT OUT
        JJ=1
        WRITE(6,165) II,JJ
        WRITE(6,166)
        CALL PRCOL4(X,CN1,Y,CN2,1,LAGH1,1)
C       CROSS COVARIANCE PUNCH OUT
        WRITE(7,1) II,JJ
        WRITE(7,2) (X(I),I=1,LAGH1)
        WRITE(7,1) JJ,II
        WRITE(7,2) (Y(I),I=1,LAGH1)
  300 STOP
    1 FORMAT(10I5)
    2 FORMAT(4D20.10)
   50 FORMAT(1H1,71HPROGRAM 5.1.3    AUTO AND/OR CROSS COVARIANCES AND CO
       ARRELATIONS VIA FFT.)
   51 FORMAT(1H0,17HINITIAL CONDITION)
   52 FORMAT(1H0,4HISW=,I5,5X,3HLD=,I5,5X,5HLAGH=,I5)
   53 FORMAT(1H0,13HORIGIANL DATA)
   54 FORMAT(1H0,4X,1HI,12X,4HX(I))
   55 FORMAT(1H0,4X,1HI,12X,4HX(I),10X,4HY(I))
  162 FORMAT(//1H0,14HAUTOCOVARIANCE,5X,6HCIJ(L),5X,2HI=,I5,5X,2HJ=,I5,5
       AX,5HMEAN=,D15.5)
  163 FORMAT(1H0,4X,1HL,5X,6HCIJ(L),8X,10HNORMALIZED)
  165 FORMAT(//1H0,16HCROSS COVARIANCE,5X,6HCIJ(L),5X,2HI=,I5,5X,2HJ=,I5
       A)
  166 FORMAT(1H0,4X,1HL,5X,6HCIJ(L),8X,10HNORMALIZED,4X,6HCJI(L),8X,10HN
       AORMALIZED)
        END
```

```
      SUBROUTINE DMEADL(X,N,XMEAN)
C     DOUBLE PRECISION MEAN DELETION
      DOUBLE PRECISION X,XMEAN,AN
      DIMENSION X(N)
      AN=N
      XMEAN=DSUMF(X,N)/AN
      DO 10 I=1,N
   10 X(I)=X(I)-XMEAN
      RETURN
      END
C
      DOUBLE PRECISION FUNCTION DSUMF(X,N)
C     DOUBLE PRECISION SUMMATION
      DOUBLE PRECISION X
      DIMENSION X(N)
      DSUMF=0.0
      DO 10 I=1,N
   10 DSUMF=DSUMF+X(I)
      RETURN
      END
C
      SUBROUTINE MIXRAD(Z,N,N2P,ISG)
C     MIXED RADIX FAST FOURIER TRANSFORM
C     ISG=-1...FOURIER TRANSFORM
C     ISG=1...INVERSE FOURIER TRANSFORM
      IMPLICIT REAL*8(A-H,O-Y)
      IMPLICIT COMPLEX*16(Z)
C     N=2**N2P (N2P IS LESS THAN 12.)
      DIMENSION Z(N)
      DIMENSION MS(11)
      AN=N
      PI=3.1415926536
      PI2=PI+PI
      SG=ISG
      ZCI=SG*(0.0,1.0)
      DO 10 I=1,N2P
   10 MS(I)=2**(N2P-I)
      N3=N2P/2
      M=N
```

```
      DO 11 L=1,N3
      M=M/4
      M4=M*4
      LM4=N-M4+1
      AM4=M4
      AM5=SG*PI2/AM4
      DO 12 J=1,M
      JM1=J-1
      AJM1=JM1
      ARG=AJM1*AM5
      C1=DCOS(ARG)
      S1=DSIN(ARG)
      C2=C1*C1-S1*S1
      S2=C1*S1+C1*S1
      C3=C1*C2-S1*S2
      S3=C1*S2+C2*S1
      ZW1=DCMPLX(C1,S1)
      ZW2=DCMPLX(C2,S2)
      ZW3=DCMPLX(C3,S3)
      DO 13 I=1,LM4,M4
      J1=I+JM1
      J2=J1+M
      J3=J2+M
      J4=J3+M
      ZC1=Z(J1)+Z(J3)
      ZC2=Z(J1)-Z(J3)
      ZC3=Z(J2)+Z(J4)
      ZC4=Z(J2)-Z(J4)
      Z(J1)=ZC1+ZC3
      Z(J2)=(ZC1-ZC3)*ZW2
      ZC4=ZCI*ZC4
      Z(J3)=(ZC2+ZC4)*ZW1
      Z(J4)=(ZC2-ZC4)*ZW3
   13 CONTINUE
   12 CONTINUE
   11 CONTINUE
      N5=N2P-2*N3
      IF(N5.NE.1) GO TO 120
      NM1=N-1
      DO 110 I=1,NM1,2
      I1=I+1
      ZC=Z(I)+Z(I1)
      Z(I1)=Z(I)-Z(I1)
      Z(I)=ZC
  110 CONTINUE
C     UNSCRAMBLING
  120 JF=0
      DO 16 I=1,N
      IF(JF.LT.I) GO TO 17
      ZC=Z(I)
      Z(I)=Z(JF+1)
      Z(JF+1)=ZC
   17 DO 18 L=1,N2P
      LL=L
      IF(JF.LT.MS(L)) GO TO 19
   18 JF=JF-MS(L)
   19 JF=JF+MS(LL)
   16 CONTINUE
      IF(ISG.LT.0) GO TO 30
      DO 20 I=1,N
   20 Z(I)=Z(I)/AN
   30 RETURN
      END
```

5.2 **Computation of Spectra**

The estimate of a spectral density function is obtained by a proper smoothing of the Fourier transform of the covariance function. The Goertzel method of Fourier transformation is useful for this purpose [2]. The Goertzel method can be regarded as a digital sine wave generator defined with a parameter that determines the frequency. Since this generator has a linear structure, it can easily be seen that if the input $G(I)$ $(I=0, 1, ...)$ is given to the system in reverse order of time, the output of the system when the input is completed gives the desired Fourier transform. Both the sine and cosine transforms can be obtained simultaneously.

[Programs]
  5.2.1 AUSPEC Power Spectrum Computation (uni-variate)
  5.2.2 MULSPE Cross Spectrum Computation (multi-variate)
  5.2.3 SGLFRE Frequency Response Function Computation (single-input)
  5.2.4 MULFRE Frequency Response Function Computation (multi-input)

```
      PROGRAM AUSPEC
C     PROGRAM 5.2.1    POWER SPECTRUM
C     THIS PROGRAM COMPUTES POWER SPECTRUM ESTIMATES FOR TWO
C     TRIGONOMETRIC WINDOWS OF BLACKMAN-TUKEY TYPE BY GOERTZEL METHOD.
C     ONLY ONE CARD OF LAGH(MAXIMUM LAG OF COVARIANCES TO BE USED FOR
C     POWER SPECTRUM COMPUTATION) SHOULD BE ADDED ON TOP OF THE OUTPUT
C     OF PROGRAM 5.1.1 AUTCOR TO FORM INPUT TO THIS PROGRAM.
C     OUTPUTS ARE ESTIMATES P1(I),P2(I) FOR FREQUENCIES I/(2LAGH*DELTAT)
C     AND THE TEST STATISTICS Q(I) FOR THE DIFFERENCES BETWEEN P1(I) AND
C     P2(I).   Q(I) GREATER THAN 1 MEANS SIGNIFICANT DIFFERENCE.
      IMPLICIT REAL*8(A-H,O-Z)
      DIMENSION CXX(501),FC(501),P1(501),P2(501),Q(501)
      DIMENSION A1(10),A2(10)
C     WINDOW W1 DEFINITION
      MLA1=2
      A1(1)=0.5
      A1(2)=0.25
C     WINDOW W2 DEFINITION
      MLA2=3
      A2(1)=0.625
      A2(2)=0.25
      A2(3)=-0.0625
C     LAGH SPECIFICATION
      READ(5,1) LAGH
      LAGH1=LAGH+1
C     READING THE OUTPUTS OF PROGRAM 5.1.1 AUTCOR
      READ(5,1) N,LAGH0
      LAGH3=LAGH0+1
C     INITIAL CONDITION PRINT OUT
      WRITE(6,60)
      WRITE(6,61)
      WRITE(6,62) N,LAGH
      WRITE(6,63)
      CALL PRCOL1(A1,1,MLA1,1)
      WRITE(6,64)
      CALL PRCOL1(A2,1,MLA2,1)
C     AUTO COVARIANCE INPUT
      READ(5,2) (CXX(I),I=1,LAGH3)
C     AUTO COVARIANCE PRINT OUT
      WRITE(6,172)
      WRITE(6,162)
      CALL PRCOL1(CXX,1,LAGH1,1)
      DO 10 I=2,LAGH
   10 CXX(I)=CXX(I)+CXX(I)
C     F-COS TRANSFORMATION
      CALL FGERCO(CXX,LAGH1,FC,LAGH1)
C     SPECTRUM SMOOTHING BY WINDOW W1
      CALL AUSP(FC,P1,LAGH1,A1,MLA1)
C     SPECTRUM SMOOTHING BY WINDOW W2
      CALL AUSP(FC,P2,LAGH1,A2,MLA2)
C     TEST STATISTICS COMPUTATION
      CALL SIGNIF(P1,P2,Q,LAGH1,N)
C     AUTO SPECTRUM AND TEST STATISTICS PRINT OUT
      WRITE(6,66) N,LAGH
      WRITE(6,67)
      CALL PRCOL3(P1,P2,Q,1,LAGH1,1)
C     AUTO SPECTRUM SMOOTHED BY WINDOW W1 PUNCH OUT
      WRITE(7,1) N,LAGH
      WRITE(7,2) (P1(I),I=1,LAGH1)
      STOP
    1 FORMAT(10I5)
    2 FORMAT(4D20.10)
   60 FORMAT(1H1,30HPROGRAM 5.2.1    POWER SPECTRUM)
   62 FORMAT(1H0,2HN=,I5,5X,5HLAGH=,I5)
   61 FORMAT(1H0,1/HINITIAL CONDITION)
   63 FORMAT(1H0,12X,9HWINDOW W1/1H ,4X,1HI,11X,5HA1(I))
   64 FORMAT(1H0,12X,9HWINDOW W2/1H ,4X,1HI,11X,5HA2(I))
   66 FORMAT(//1H0,14HPOWER SPECTRUM,5X,2HN=,I5,5X,5HLAGH=,I5)
   67 FORMAT(1H ,4X,1HI,8X,8HPOWER W1,6X,8HPOWER W2,2X,12HSIGNIFICANCE)
  162 FORMAT(1H0,4X,1HI,6X,6HCXX(I))
  172 FORMAT(1H0,5X,15HAUTO COVARIANCE)
      END
```

```
      PROGRAM MULSPE
C     PROGRAM 5.2.2    MULTIPLE SPECTRUM
C     THIS PROGRAM COMPUTES MULTIPLE SPECTRUM ESTIMATES FROM THE OUTPUT
C     OF PROGRAM 5.1.2 MULCOR, USING WINDOWS W1 AND W2.
C     ONLY ONE CARD OF LAGH(MAXIMUM LAG OF COVARIANCES TO BE USED FOR
C     SPECTRUM COMPUTATION) SHOULD BE ADDED ON TOP OF THE OUTPUT OF
C     PROGRAM 5.1.2 MULCOR TO FORM THE INPUT TO THIS PROGRAM.
C     IN THE CARD OUTPUT OF SPECTRUM MATRIX ON AND LOWER DIAGONAL ARE
C     REAL PARTS AND UPPER DIAGONAL ARE IMAGINARY PARTS OF ON AND LOWER
C     DIAGONAL SPECTRAL ELEMENTS.
      IMPLICIT REAL*8(A-H,O-Z)
      DIMENSION C(501),S(501),G(501)
      DIMENSION FC(501),FS(501),P1(501),P2(501),P3(501)
      DIMENSION A1(10),A2(10)
      DIMENSION P(10020)
      DIMENSION X(101,10,10)
C     WINDOW W1 DEFINITION
      MLA1=2
      A1(1)=0.5
      A1(2)=0.25
C     WINDOW W2 DEFINITION
      MLA2=3
      A2(1)=0.625
      A2(2)=0.25
      A2(3)=-0.0625
C     LAGH SPECIFICATION
      READ(5,1) LAGH
      LAGH1=LAGH+1
C     READING THE OUTPUTS OF PROGRAM 5.1.2 MULCOR
      READ(5,1) N,LAGH0,K
      LAGH3=LAGH0+1
C     INITIAL CONDITION PRINT AND PUNCH OUT
      WRITE(6,60)
      WRITE(6,61)
      WRITE(6,62) N,LAGH,K
      WRITE(6,63)
      CALL PRCOL1(A1,1,MLA1,1)
      WRITE(6,64)
      CALL PRCOL1(A2,1,MLA2,1)
      WRITE(7,1) N,LAGH,K
C     COMPUTATION STARTS HERE WITH DISK AS EXTERNAL MEMORY.
      REWIND 1
      DO 10 II=1,K
C     AUTO COVARIANCE INPUT
      READ(5,1) IR0,IC0
      READ(5,2) (C(I),I=1,LAGH3)
      DO 20 I=1,LAGH1
   20 G(I)=C(I)+C(I)
      G(1)=0.5*G(1)
      G(LAGH1)=0.5*G(LAGH1)
C     F-COS TRANSFORMATION
      CALL FGERCO(G,LAGH1,FC,LAGH1)
C     SPECTRUM SMOOTHING BY WINDOW W1
      CALL AUSP(FC,P1,LAGH1,A1,MLA1)
C     SPECTRUM SMOOTHING BY WINDOW W2
      CALL AUSP(FC,P2,LAGH1,A2,MLA2)
C     TEST STATISTICS COMPUTATION
      CALL SIGNIF(P1,P2,P3,LAGH1,N)
C     AUTO SPECTRUM AND TEST STATISTICS PRINT OUT
      WRITE(6,65) IR0,IC0
      WRITE(6,66)
      WRITE(6,67)
      CALL PRCOL3(P1,P2,P3,1,LAGH1,1)
```

```
C       AUTO SPECTRUM STORE
        LAG2=LAGH1+LAGH1
        III1=(II-1)*LAG2
        DO 21 I=1,LAGH1
        I1=III1+I
        I2=I1+LAGH1
        P(I1)=P1(I)
   21   P(I2)=P2(I)
        IF(II.EQ.1) GO TO 10
C       CROSS COVARIANCE INPUT
        IM1=II-1
        DO 11 JJ=1,IM1
        READ(5,1) IR1,IC1
        READ(5,2) (C(I),I=1,LAGH3)
        READ(5,1) IR2,IC2
        READ(5,2) (S(I),I=1,LAGH3)
C       F-COS TRANSFORMATION
        DO 30 I=1,LAGH1
   30   G(I)=C(I)+S(I)
        G(1)=0.5*G(1)
        G(LAGH1)=0.5*G(LAGH1)
        CALL FGERCO(G,LAGH1,FC,LAGH1)
C       F-SIN TRANSFORMATION
        DO 31 I=1,LAGH1
   31   G(I)=S(I)-C(I)
        G(1)=0.5*G(1)
        G(LAGH1)=0.5*G(LAGH1)
        CALL FGERSI(G,LAGH1,FS,LAGH1)
C       SMOOTHING BY WINDOW W1
        ISW=1
        IWD=0
        CALL CROSSP(FC,FS,P1,P2,LAGH1,A1,MLA1)
C       SIMPLE COHERENCE COMPUTATION
   33   III1=(II-1)*LAG2+IWD
        JJ1=(JJ-1)*LAG2+IWD
        DO 32 I=1,LAGH1
        I1=III1+I
        I2=JJ1+I
        C(I)=P(I1)
   32   S(I)=P(I2)
        CALL SIMCOH(P1,P2,C,S,P3,LAGH1)
C       CROSS SPECTRUM AND SIMPLE COHERENCE   PRINT OUT
        WRITE(6,65) IR1,IC1
        IF(ISW.NE.1) GO TO 260
        WRITE(6,166)
        GO TO 268
  260   WRITE(6,266)
  268   WRITE(6,167)
        CALL PRCOL3(P1,P2,P3,1,LAGH1,1)
        IF(ISW.LT.0) GO TO 11
C       CROSS SPECTRUM STORE (DISK)
        WRITE(1) (P1(I),I=1,LAGH1),(P2(I),I=1,LAGH1)
C       SMOOTHING BY WINDOW W2
        ISW=-1
        IWD=LAGH1
        CALL CROSSP(FC,FS,P1,P2,LAGH1,A2,MLA2)
        GO TO 33
   11   CONTINUE
   10   CONTINUE
        END FILE 1
```

```
C       SPECTRUM (SMOOTHED BY WINDOW W1) PUNCH OUT
        LC=LAGH1
        IL=101
        ILM1=IL-1
        J1=0
        J2=0
        IB=0
  416   J1=J2+1
        J2=J1+ILM1
        IF(J2.LE.LC) GO TO 413
  412   J2=LC
  413   REWIND 1
        DO 500 II=1,K
        II1=(II-1)*LAG2
        DO 520 I=J1,J2
        I0=I-IB
        I1=II1+I
  520   X(I0,II,II)=P(I1)
        IF(II.EQ.1) GO TO 500
        IM1=II-1
        DO 501 JJ=1,IM1
        READ(1) (P1(I),I=1,LAGH1),(P2(I),I=1,LAGH1)
        DO 530 I=J1,J2
        I0=I-IB
        X(I0,II,JJ)=P1(I)
  530   X(I0,JJ,II)=P2(I)
  501   CONTINUE
  500   CONTINUE
        DO 610 I=J1,J2
        I0=I-IB
        DO 611 II=1,K
  611   WRITE(7,2) (X(I0,II,JJ),JJ=1,K)
  610   CONTINUE
  417   IB=IB+IL
        IF(J2.LT.LC) GO TO 416
        STOP
    1   FORMAT(10I5)
    2   FORMAT(4D20.10)
   60   FORMAT(1H1,13HPROGRAM 5.2.2,3X,17HMULTIPLE SPECTRUM)
   61   FORMAT(1H0,17HINITIAL CONDITION)
   62   FORMAT(1H0,2HN=,I5,5X,5HLAGH=,I5,5X,2HK=,I5)
   63   FORMAT(1H0,12X,9HWINDOW W1/1H ,4X,1HI,11X,5HA1(I))
   64   FORMAT(1H0,12X,9HWINDOW W2/1H ,4X,1HI,11X,5HA2(I))
   65   FORMAT(//1H0,8HP(II,JJ),5X,3HII=,I5,3X,3HJJ=,I5)
   66   FORMAT(1H0,7X,14HPOWER SPECTRUM)
   67   FORMAT(1H ,4X,1HI,8X,8HPOWER W1,6X,8HPOWER W2,2X,12HSIGNIFICANCE)
  166   FORMAT(1H0,7X,14HCROSS SPECTRUM,8X,2HW1)
  266   FORMAT(1H0,7X,14HCROSS SPECTRUM,8X,2HW2)
  167   FORMAT(1H ,4X,1HI,5X,11HCO-SPECTRUM,1X,13HQUAD-SPECTRUM,2X,16HSIMP
       ALE COHERENCE)
        END
```

```
      SUBROUTINE CRUSSP(FC,FS,P1,P2,LAGH1,A,LA1)
C     THIS SUBROUTINE COMPUTES SMOOTHED CROSS SPECTRUM.
C     FC,FS: OUTPUTS OF FGERCO AND FGERSI
C     P1,P2: REAL AND IMAGINARY PART OF SMOOTHED CROSS SPECTRUM
C     LAGH1: DIMENSION OF FC, FS AND PI (I=1,2)
C     A: SMOOTHING COEFFICIENTS
C     LA1: DIMENSION OF A (LESS THAN 11)
      IMPLICIT REAL*8(A-H,O-Z)
      DIMENSION FC(LAGH1),FS(LAGH1),P1(LAGH1),P2(LAGH1)
      DIMENSION A(LA1)
      DIMENSION FC1(521),FS1(521)
      LA=LA1-1
      LAGSHF=LAGH1+2*LA
C     FC SHIFT-RIGHT BY LA FOR END CORRECTION
      CALL ECORCO(FC,LAGH1,FC1,LAGSHF,LA1)
C     REAL PART SMOOTHING
      CALL SMOSPE(FC1,LAGSHF,A,LA1,P1,LAGH1)
C     FS SHIFT-RIGHT BY LA FOR END CORRECTION
      CALL ECORSI(FS,LAGH1,FS1,LAGSHF,LA1)
C     IMAGINARY PART SMOOTHING
      CALL SMOSPE(FS1,LAGSHF,A,LA1,P2,LAGH1)
      RETURN
      END
C
      SUBROUTINE ECORSI(FS,LAGH1,FS1,LAGSHF,LA1)
C     FS SHIFT-RIGHT BY LA FOR IMAGINARY PART END CORRECTION
      IMPLICIT REAL*8(A-H,O-Z)
      DIMENSION FS(LAGH1),FS1(LAGSHF)
      LAGH2=LAGH1+1
      LA=LA1-1
      DO 100 I=1,LAGH1
      I1=LAGH2-I
      I2=I1+LA
  100 FS1(I2)=FS(I1)
      LA2=LAGH1+LA
      DO 110 I=1,LA
      I1=LA1-I
      I2=LA1+I
      I3=LA2-I
      I4=LA2+I
      FS1(I1)=-FS1(I2)
  110 FS1(I4)=-FS1(I3)
      RETURN
      END
```

```
      SUBROUTINE FGERSI(G,LGP1,FS,LF1)
C     FOURIER TRANSFORM (GOERTZEL METHOD)
C     THIS SUBROUTINE COMPUTES FOURIER TRANSFORM OF G(I),I=0,1,...,LG AT
C     FREQUENCIES K/(2*LF),K=0,1,...,LF AND RETURNS SIN TRANSFORM IN
C     FS(K).
      IMPLICIT REAL*8(A-H,O-Z)
      DIMENSION G(LGP1),FS(LF1)
      LG=LGP1-1
      LF=LF1-1
C     REVERSAL OF G(I),I=1,...,LGP1 INTO G(LG3-I)    LG3=LGP1+1
      IF(LGP1.LE.1) GO TO 110
      LG3=LGP1+1
      LG4=LGP1/2
      DO 100 I=1,LG4
      I2=LG3-I
      T=G(I)
      G(I)=G(I2)
  100 G(I2)=T
  110 PI=3.1415926536
      ALF=LF
      T=PI/ALF
      DO 10 K=1,LF1
      AK=K-1
      TK=T*AK
      CK=DCOS(TK)
      SK=DSIN(TK)
      CK2=CK+CK
      UM2=0.0
      UM1=0.0
      IF(LG.EQ.0) GO TO 12
      DO 11 I=1,LG
      UM0=CK2*UM1-UM2+G(I)
      UM2=UM1
   11 UM1=UM0
   12 FS(K)=SK*UM1
   10 CONTINUE
      RETURN
      END
C
      SUBROUTINE SIMCOH(P1,P2,C,S,P3,LAGH1)
C     THIS SUBROUTINE COMPUTES SIMPLE COHERENCE.
      IMPLICIT REAL*8(A-H,O-Z)
      DIMENSION P1(LAGH1),P2(LAGH1),C(LAGH1),S(LAGH1),P3(LAGH1)
      DO 10 I=1,LAGH1
   10 P3(I)=(P1(I)**2+P2(I)**2)/(C(I)*S(I))
      RETURN
      END
```

```
      PROGRAM SGLFRF
C     PROGRAM 5.2.3   FREQUENCY RESPONSE FUNCTION (SINGLE CHANNEL)
C     THIS PROGRAM COMPUTES 1-INPUT,1-OUTPUT FREQUECNY RESPONSE FUNCTION
C     ,GAIN,PHASE,COHERENCY AND RELATIVE ERROR STATISTICS.
C     ONE CARD WITH SPECIFICATION OF INPUT(INP) AND OUTPUT(IOUT)
C     VARIABLE SHOULD BE ADDED ON TOP OF THE OUTPUT OF PROGRAM 5.2.2
C     MULSPE TO FORM THE INPUT TO THIS PROGRAM.
C     WITHIN IP VARIABLES OF MULSPE OUTPUT, INP-TH AND IOUT-TH VARIABLE
C     ARE TAKEN AS INPUT AND OUTPUT VARIABLE.
      IMPLICIT REAL*8(A-H,O-Z)
      DIMENSION P11(501),P22(501),C(501),S(501)
      DIMENSION R(501),PH(501),P(10,10)
C     ABSOLUTE DIMENSION USED FOR SUBROUTINE CALL
      MJ=10
C     INPUT OUTPUT VARIABLE SPECIFICATION
      READ(5,1) INP,IOUT
C     READING THE OUTPUT OF PROGRAM 5.2.2 MULSPE
      READ(5,1) N,LAGH,IP
      LAGH1=LAGH+1
      DO 5 I=1,LAGH1
      CALL REMATX(P,IP,IP,1,MJ,MJ)
C     MATRIX REARRANGEMENT
      P11(I)=P(INP,INP)
      P22(I)=P(IOUT,IOUT)
      IF(INP.LT.IOUT) GO TO 7
      C(I)=P(INP,IOUT)
      S(I)=-P(IOUT,INP)
      GO TO 5
    7 C(I)=P(IOUT,INP)
      S(I)=P(INP,IOUT)
    5 CONTINUE
C     INITIAL CONDITION PRINT OUT
      WRITE(6,55)
      WRITE(6,56)
      WRITE(6,57) N,LAGH
      WRITE(6,259) INP,IOUT
      WRITE(6,58)
      WRITE(6,59)
      WRITE(6,159)
      CALL PRCOL4(P11,P22,C,S,1,LAGH1,1)
C     FREQUENCY RESPONSE FUNCTION COMPUTATION
      DO 10 I=1,LAGH1
      C(I)=C(I)/P11(I)
      S(I)=S(I)/P11(I)
   10 P22(I)=P22(I)/P11(I)
C     GAIN COMPUTATION
      DO 11 I=1,LAGH1
      R(I)=C(I)**2+S(I)**2
   11 P11(I)=DSQRT(R(I))
C     PHASE COMPUTATION
      CALL SPHASE(C,S,PH,LAGH1)
C     COHERENCY COMPUTATION
      DO 12 I=1,LAGH1
   12 P22(I)=R(I)/P22(I)
C     RELATIVE ERROR STATISTICS COMPUTATION
      CALL SGLERR(P22,R,N,LAGH1)
```

```
C       FREQUENCY RESPONSE FUNCTION, GAIN, PHASE, COHERENCY AND RELATIVE
C       ERROR STATISTICS PRINT OUT
        WRITE(6,60)
        WRITE(6,61)
        CALL PRCOL6(C,S,P11,PH,P22,R,1,LAGH1,1)
        STOP
    1 FORMAT(10I5)
   55 FORMAT(1H1,60HPROGRAM 5.2.3    FREQUENCY RESPONSE FUNCTION (SINGLE
      ACHANNEL))
   56 FORMAT(1H0,17HINITIAL CONDITION)
   57 FORMAT(1H0,2HN=,I5,5X,5HLAGH=,I5)
   58 FORMAT(1H0,22HINITIAL DATA(SPECTRUM))
   59 FORMAT(1H0,6X,14HPOWER SPECTRUM,1X,14HPOWER SPECTRUM,3X,11HCO-SPEC
      ATRUM,1X,13HQUAD-SPECTRUM)
  159 FORMAT(1H ,4X,1HI,10X,6HP(1,1),8X,6HP(2,2),10X,4HC(I),10X,4HS(I))
   69 FORMAT(//1H0,4X,1HI,3X,27HFREQUENCY RESPONSE FUNCTION,10X,4HGAIN,9
      AX,5HPHASE,5X,9HCOHERENCY,6X,8HRELATIVE)
   61 FORMAT(1H ,12X,9HREAL PART,4X,10HIMAG. PART,51X,5HERROR)
  259 FORMAT(1H0,6HINPUT=,I5,5X,7HOUTPUT=,I5)
        END
C
        SUBROUTINE SGLARC(C,S,ARC,LAGH1)
C       THIS SUBROUTINE COMPUTES RAW PHASES.
C       (SINGLE CHANNEL)
        IMPLICIT REAL*8(A-H,O-Z)
        DIMENSION C(LAGH1),S(LAGH1),ARC(LAGH1)
        PI=3.1415926536
        DO 10 I=1,LAGH1
        IF(C(I)) 11,12,13
   11 IF(S(I)) 14,15,16
   12 IF(S(I)) 17,18,19
   13 ARC(I)=DATAN(S(I)/C(I))
        GO TO 10
   14 ARC(I)=DATAN(S(I)/C(I))-PI
        GO TO 10
   15 ARC(I)=-PI
        GO TO 10
   16 ARC(I)=DATAN(S(I)/C(I))+PI
        GO TO 10
   17 ARC(I)=-PI/2.0
        GO TO 10
   18 ARC(I)=0.0
        GO TO 10
   19 ARC(I)=PI/2.0
   10 CONTINUE
        RETURN
        END
```

```
        SUBROUTINE SGLERR(CH,R,N,LAGH1)
C       THIS SUBROUTINE COMPUTES RELATIVE ERROR STATISTICS.
C       (SINGLE CHANNEL)
        IMPLICIT REAL*8(A-H,O-Z)
        DIMENSION CH(LAGH1),R(LAGH1)
C       CONSTANTS D1,D2 COMPUTATION
        LAGH=LAGH1-1
        CALL SUBD12(N,LAGH,1,D1,D2)
C       RELATIVE ERROR STATISTICS COMPUTATION
        DO 20 I=1,LAGH1
        IF(CH(I).LE.0.0) GO TO 22
        IF(CH(I).GT.1.0) GO TO 22
        E1=1.0/CH(I)-1.0
        ER=DSQRT(E1)
        IF(I.EQ.1) GO TO 23
        IF(I.EQ.LAGH1) GO TO 23
        R(I)=D2*ER
        GO TO 20
     23 R(I)=D1*ER
        GO TO 20
     22 R(I)=100.0
     20 CONTINUE
        RETURN
        END
C
        SUBROUTINE SGLPAC(ARC,PH,LAGH1)
C       THIS SUBROUTINE MAKES PHASE CURVE CONTINUOUS.
C       (SINGLE CHANNEL)
        IMPLICIT REAL*8(A-H,O-Z)
        DIMENSION ARC(LAGH1),PH(LAGH1)
        PI=3.1415926536
        PI2=PI+PI
        PH(1)=ARC(1)
        DO 10 I=2,LAGH1
        DK=ARC(I)-ARC(I-1)
        IF(DK.GT.PI) GO TO 11
        IF(DK.LT.-PI) GO TO 12
        PH(I)=PH(I-1)+DK
        GO TO 10
     11 PH(I)=PH(I-1)+DK-PI2
        GO TO 10
     12 PH(I)=PH(I-1)+DK+PI2
     10 CONTINUF
        RETURN
        END
C
        SUBROUTINE SPHASE(C,S,PH,LAGH1)
C       THIS SUBROUTINE COMPUTES PHASE.
C       (SINGLE CHANNEL)
        IMPLICIT REAL*8(A-H,O-Z)
        DIMENSION C(LAGH1),S(LAGH1),PH(LAGHI)
        DIMENSION ARC(501)
C       ARCTANGENT COMPUTATION
        CALL SGLARC(C,S,ARC,LAGH1)
C       PHASE COMPUTATION
        CALL SGLPAC(ARC,PH,LAGH1)
        RETURN
        END
```

```
      PROGRAM MULFRF
C     PROGRAM 5.2.4   FREQUENCY RESPONSE FUNCTION (MULTIPLE CHANNEL)
C     THIS PROGRAM COMPUTES MULTIPLE FREQUENCY RESPONSE FUNCTION, GAIN,
C     PHASE, MULTIPLE COHERENCY, PARTIAL COHERENCY AND RELATIVE ERROR
C     STATISTICS.
C     A CARD WITH THE TOATL NUMBER(K) OF INPUT VARIABLES AND ANOTHER
C     WITH SPECIFICATION OF INPUT VARIABLES(INW(I),I=1,K) AND OUTPUT
C     VARIABLE(INW(K+1)) SHOULD BE ADDED ON TOP OF THE OUTPUT OF
C     PROGRAM 5.2.2 MULSPE TO FORM THE INPUT TO THIS PROGRAM.
C     WITHIN IPO VARIABLES OF MULSPE OUTPUT, ONLY THOSE K+1 INW(I)-TH
C     VARIABLES ARE TAKEN INTO COMPUTATION.
      IMPLICIT REAL*8(A-H,O-W)
      IMPLICIT COMPLEX*16(X-Z)
      DIMENSION P(10,10),X(10,10),C(10),S(10),G(10)
      DIMENSION OARC(10),PH(10),PCH(10),R(10),INW(10)
C     ABSOLUTE DIMENSION USED FOR SUBROUTINE CALL
      MJ=10
C     INPUT OUTPUT VARIABLE SPECIFICATION
      READ(5,1) K
      K1=K+1
      READ(5,1) (INW(I),I=1,K1)
C     FOLLOWING INPUTS ARE OUTPUTS OF PROGRAM 5.2.2 MULSPE.
      READ(5,1) N,LAGH,IPO
      LAGH1=LAGH+1
C     INITIAL CONDITION PRINT OUT
      WRITE(6,55)
      WRITE(6,56)
      WRITE(6,57) N,LAGH,K
      WRITE(6,259) (INW(I),I=1,K1)
C     COMPUTATION START
      DO 10 JF=1,LAGH1
      JJF=JF
      JFM1=JF-1
      WRITE(6,58) JFM1
C     SPECTRUM INPUT
      CALL REMATX(P,IPO,IPO,1,MJ,MJ)
C     REAL TO COMPLEX TRANSFORMATION
      DO 401 I=1,IPO
      X(I,I)=P(I,I)
      IF(I.EQ.1) GO TO 401
      IM1=I-1
      DO 402 J=1,IM1
      X(I,J)=DCMPLX(P(I,J),P(J,I))
  402 X(J,I)=DCONJG(X(I,J))
  401 CONTINUE
C     MATRIX REARRANGEMENT AND PRINT OUT (COMPLEX)
      CALL REARRA(X,INW,IPO,K1,MJ)
      WRITE(6,159)
      CALL PRCPMA(X,K1,K1,MJ,MJ)
C     FREQUENCY RESPONSE FUNCTION COMPUTATION
      P00=DREAL(X(K1,K1))
      CALL FQCPIV(X,XDET,K,MJ)
      DO 20 I=1,K
      C(I)=DREAL(X(I,K1))
   20 S(I)=-DIMAG(X(I,K1))
C     GAIN COMPUTATION
      DO 21 I=1,K
   21 G(I)=DSQRT(C(I)**2+S(I)**2)
C     PHASE COMPUTATION
      CALL MPHASE(C,S,OARC,PH,K,JJF)
```

```
C       PARTIAL COHERENCY AND MULTIPLE COHERENCY COMPUTATION
        EP=DREAL(X(K1,K1))
        DO 22 I=1,K
        G2=G(I)**2
        G3=G2+EP*X(I,I)
        IF(G3.NE.0.0) GO TO 23
        PCH(I)=100.0
        GO TO 22
     23 PCH(I)=G2/G3
     22 CONTINUE
        CHM=1.0-EP/P00
C       RELATIVE ERROR STATISTICS COMPUTATION
        CALL MULERR(PCH,R,N,LAGH1,K,JJF,D1,D2)
C       FREQUENCY RESPONSE FUNCTION, GAIN, PHASE, PARTIAL COHERENCY,
C       MULTIPLE COHERENCY, RELATIVE ERROR STATISTICS PRINT OUT
        WRITE(6,60)
        WRITE(6,61)
        CALL PRCOL6(C,S,G,PH,PCH,R,1,K,0)
        WRITE(6,65) CHM
     10 CONTINUE
        STOP
      1 FORMAT(10I5)
     55 FORMAT(1H1,62HPROGRAM 5.2.4    FREQUENCY RESPONSE FUNCTION (MULTIPL
        AE CHANNEL))
     56 FORMAT(1H0,17HINITIAL CONDITION)
     57 FORMAT(1H0,2HN=,I5,5X,5HLAGH=,I5,5X,2HK=,I5)
     58 FORMAT(//1H0,2HF=,I5)
     60 FORMAT(//1H0,4X,1HI,3X,27HFREQUENCY RESPONSE FUNCTION,10X,4HGAIN,9
        AX,5HPHASE,7X,7HPARTIAL,6X,8HRELATIVE)
     61 FORMAT(1H ,12X,9HREAL PART,4X,10HIMAG. PART,33X,9HCOHERENCY,9X,5HE
        ARROR)
     65 FORMAT(1H ,69X,8HMULTIPLE/1H ,68X,9HCOHERENCY/1H ,63X,D14.5)
    159 FORMAT(1H0,28HSPECTRUM MATRIX (REARRANGED))
    259 FORMAT(/1H0,6HINW(I),5X,10I5)
        END
```

```
      SUBROUTINE FQCPIV(X,XDET,MM,MJ)
C     THIS SUBROUTINE COMPUTES MULTIPLE FREQUENCY RESPONSE FUNCTION.
C     MM: THE TOTAL NUMBER OF INPUTS (LESS THAN 10)
C     MJ: ABSOLUTE DIMENSION OF X IN THE MAIN ROUTINE
      IMPLICIT COMPLEX*16(X)
      DIMENSION X(MJ,MJ)
      DIMENSION IDS(10)
      XDET=1.0
      MP1=MM+1
      DO 10 L=1,MM
C     PIVOTING AT L-TH STAGE
      XMAXP=0.10000D-10
      MAXI=0
      DO 110 I=L,MM
      IF(CDABS(XMAXP).GE.CDABS(X(I,L))) GO TO 110
      XMAXP=X(I,L)
      MAXI=I
  110 CONTINUE
      IDS(L)=MAXI
      IF(MAXI.EQ.L) GO TO 120
      IF(MAXI.GT.0) GO TO 121
      XDET=0.0
      GO TO 140
C     ROW INTERCHANGE
  121 DO 14 J=1,MP1
      XC=X(MAXI,J)
      X(MAXI,J)=X(L,J)
   14 X(L,J)=XC
      XDET=-XDET
  120 XDET=XDET*XMAXP
      XC=1.0/XMAXP
      X(L,L)=1.0
      DO 11 J=1,MP1
   11 X(L,J)=X(L,J)*XC
      DO 12 I=1,MP1
      IF(I.EQ.L) GO TO 12
      XC=X(I,L)
      X(I,L)=0.0
      DO 13 J=1,MP1
   13 X(I,J)=X(I,J)-XC*X(L,J)
   12 CONTINUE
   10 CONTINUE
      IF(MM.GT.1) GO TO 123
      GO TO 140
C     COLUMN INTERCHANGE
  123 MM1=MM-1
      DO 130 J=1,MM1
      MMJ=MM-J
      JJ=IDS(MMJ)
      IF(JJ.EQ.MMJ) GO TO 130
      DO 131 I=1,MP1
      XC=X(I,JJ)
      X(I,JJ)=X(I,MMJ)
  131 X(I,MMJ)=XC
  130 CONTINUE
  140 RETURN
      END
```

```
      SUBROUTINE MPHASE(C,S,OARC,PH,K,JJF)
C     THIS SUBROUTINE COMPUTES PHASE
C     (MULTIPLE CHANNEL)
      IMPLICIT REAL*8(A-H,O-Z)
      DIMENSION C(K),S(K),OARC(K),PH(K)
      DIMENSION ARC(10)
C     ARCTANGENT COMPUTATION
      CALL MULARC(C,S,ARC,K)
C     PHASE COMPUTATION
      CALL MULPAC(ARC,OARC,PH,K,JJF)
      RETURN
      END
C
      SUBROUTINE MULARC(C,S,ARC,K)
C     THIS SUBROUTINE COMPUTES RAW PHASE.
C     (MULTIPLE CHANNEL)
      IMPLICIT REAL*8(A-H,O-Z)
      DIMENSION C(K),S(K),ARC(K)
      PI=3.1415926536
      DO 10 I=1,K
      IF(C(I)) 11,12,13
   11 IF(S(I)) 14,15,16
   12 IF(S(I)) 17,18,19
   13 ARC(I)=DATAN(S(I)/C(I))
      GO TO 10
   14 ARC(I)=DATAN(S(I)/C(I))-PI
      GO TO 10
   15 ARC(I)=-PI
      GO TO 10
   16 ARC(I)=DATAN(S(I)/C(I))+PI
      GO TO 10
   17 ARC(I)=-PI/2.0
      GO TO 10
   18 ARC(I)=0.0
      GO TO 10
   19 ARC(I)=PI/2.0
   10 CONTINUE
      RETURN
      END
C
      SUBROUTINE MULERR(PCH,R,N,LAGH1,K,JJF,D1,D2)
C     THIS SUBROUTINE COMPUTES RELATIVE ERROR STATISTICS.
C     (MULTIPLE CHANNEL)
      IMPLICIT REAL*8(A-H,O-Z)
      DIMENSION PCH(K),R(K)
      IF(JJF.NE.1) GO TO 30
C     CONSTANTS D1,D2 COMPUTATION
      LAGH=LAGH1-1
      CALL SUBD12(N,LAGH,K,D1,D2)
C     RELATIVE ERROR STATISTICS COMPUTATION
   30 DO 20 I=1,K
      IF(PCH(I).LE.0.0) GO TO 22
      IF(PCH(I).GT.1.0) GO TO 22
      E1=1.0/PCH(I)-1.0
      ER=DSQRT(E1)
      IF(JJF.EQ.1) GO TO 23
      IF(JJF.EQ.LAGH1) GO TO 23
      R(I)=D2*ER
      GO TO 20
   23 R(I)=D1*ER
      GO TO 20
   22 R(I)=100.0
   20 CONTINUE
      RETURN
      END
```

```
      SUBROUTINE MULPAC(ARC,OARC,PH,K,JJF)
C     THIS SUBROUTINE MAKES PHASE CURVE CONTINUOUS.
C     (MULTIPLE CHANNEL)
      IMPLICIT REAL*8(A-H,O-Z)
      DIMENSION ARC(K),OARC(K),PH(K)
      PI=3.1415926536
      PI2=PI+PI
      IF(JJF.NE.1) GO TO 20
      DO 9 I=1,K
      PH(I)=ARC(I)
    9 OARC(I)=ARC(I)
      GO TO 30
   20 DO 10 I=1,K
      DK=ARC(I)-OARC(I)
      IF(DK.GT.PI) GO TO 11
      IF(DK.LT.-PI) GO TO 12
      PH(I)=PH(I)+DK
      GO TO 10
   11 PH(I)=PH(I)+DK-PI2
      GO TO 10
   12 PH(I)=PH(I)+DK+PI2
   10 OARC(I)=ARC(I)
   30 RETURN
      END
```

5.3 Autoregressive Model Computation

To fit an autoregressive model, the FPE (uni-variate) or the FPEC (multi-variate) computation procedure is applied to the covariance function obtained in 5.1. The FPEC computation procedure is originally intended for model fitting for control. The MFPE computation for multi-variate autoregressive model fitting can be realized simply by treating all the variables as controlled variables and putting the number of manipulated variables equal to zero.

[Programs]
  5.3.1 FPEAUT FPE Computation (for uni-variate autoregressive model)
  5.3.2 FPEC FPEC Computation (for control system model or multi-variate autoregressive model)
  Program MULNOS gives the relative power contribution of each noise source to the total power of a variable at each frequency. This is realized by using the result of the MFPE calculation and assuming that the components of the one-step ahead prediction error vector are mutually uncorrelated. Program DECONV gives estimates of the coefficients of equation (8) of 3.3.2. Among the outputs of MULNOS, the differential and integrated power contributions give the quantities defined by the equations (14) and (15) of 3.3.2, respectively.

[Programs]
  5.3.3 MULNOS Relative Power Contribution Computation
  5.3.4 DECONV Impulse Response Computation

```
      PROGRAM FPEAUT
C     PROGRAM 5.3.1    FPE AUTO
C     THIS PROGRAM PERFORMS FPE(FINAL PREDICTION ERROR) COMPUTATION FOR
C     ONE-DIMENSIONAL AR-MODEL. A CARD CONTAINING THE FOLLOWING
C     INFORMATION OF L, UPPER LIMIT OF MODEL ORDER, SHOULD BE ADDED ON
C     TOP OF THE OUTPUT OF PROGRAM 5.1.1 AUTCOR TO FORM THE INPUT TO
C     THIS PROGRAM.
C     CXX(0) IS READ AS INITIAL SD.
C     THE OUTPUTS ARE THE COEFFICIENTS A(I) OF AR-PROCESS
C     X(N)=A(1)X(N-1)+...+A(M)X(N-M)+E(N)
C     AND THE VARIANCE SIGMA**2 OF E(N).
C     CHI**2 SHOWS THE SIGNIFICANCE OF PARCOR=A(M) AS A CHI-SQUARED
C     VARIABLE WITH D.F.=1.
      IMPLICIT REAL*8(A-H,O-Z)
      DIMENSION CXX(501),A(501),B(501),AO(501)
C     L SPECIFICATION
      READ(5,1) L
C     READING THE OUTPUT OF PROGRAM 5.1.1 AUTCOR
      READ(5,1) N,LAGH
      READ(5,2) SD,(CXX(I),I=1,LAGH)
C     COMPUTATION START
      AN=N
      NP1=N+1
      NM1=N-1
      ANP1=NP1
      ANM1=NM1
      OFPE=(ANP1/ANM1)*SD
      OOFPE=1.0/OFPE
      ORFPE=1.0
      OSD=SD
      MO=0
      WRITE(6,155)
      WRITE(6,156)
      WRITE(6,57) N,L
      WRITE(6,140)
      WRITE(6,141) SD
      CALL PRCOL1(CXX,1,L,0)
      WRITE(6,157)
      WRITE(6,58) OFPE
      SE=CXX(1)
      DO 400 M=1,L
      MP1=M+1
      D=SE/SD
      A(M)=D
      D2=D*D
      SD=(1.0-D2)*SD
      ANP1=NP1+M
      ANM1=NM1-M
      FPE=(ANP1/ANM1)*SD
      RFPE=FPE*OOFPE
      CHI2=D2*ANM1
      IF(M.EQ.1) GO TO 410
```

```
C      A(I) COMPUTATION
       LM=M-1
       DO 420 I=1,LM
420    A(I)=A(I)-D*B(I)
410    DO 421 I=1,M
       IM=MP1-I
421    B(I)=A(IM)
       WRITE(6,60) M
       WRITE(6,61) SD,FPE,RFPE
       WRITE(6,62) D,CHI2
       WRITE(6,160)
       CALL PRCOL1(A,1,M,0)
       IF(OFPE.LT.FPE) GO TO 440
       OFPE=FPE
       ORFPE=RFPE
       OSD=SD
       MO=M
       DO 430 I=1,M
430    AO(I)=A(I)
440    IF(M.EQ.L) GO TO 400
       SE=CXX(MP1)
       DO 441 I=1,M
441    SE=SE-B(I)*CXX(I)
400    CONTINUE
       WRITE(6,63) OFPE,MO
       WRITE(6,64) ORFPE
       WRITE(7,1) N,MO
       WRITE(7,2) OSD
       IF(MO.LE.0) GO TO 699
       WRITE(7,2) (AO(I),I=1,MO)
699    STOP
  1    FORMAT(10I5)
  2    FORMAT(4D20.10)
 57    FORMAT(1H0,2HN=,I5,5X,2HL=,I5)
 58    FORMAT(1H0,5HOFPE=,D12.5)
 60    FORMAT(1H0,2HM=,I5)
 61    FORMAT(1H ,9HSIGMA**2=,D12.5,2X,4HFPE=,D12.5,2X,5HRFPE=,D12.5)
 62    FORMAT(1H0,7HPARCOR=,D14.5,2X,15HCIH**2(D.F.=1)=,D12.5)
160    FORMAT(1H0,4X,1HI,12X,4HA(I))
 63    FORMAT(1H0,13HMINIMUM FPE =,D12.5,2X,14HATTAINED AT M=,I5)
 64    FORMAT(1H ,13HMINIMUM RFPE=,D12.5)
140    FORMAT(1H0,4X,1HI,5X,15HAUTO COVARIANCE)
141    FORMAT(1H ,4X,1H0,D16.5)
155    FORMAT(1H1,24HPROGRAM 5.3.1    FPE AUTO
156    FORMAT(1H0,17HINITIAL CONDITION)
157    FORMAT(1H0,2HM=,4X,1H0)
       END
```

```
        PROGRAM FPEC
C       PROGRAM 5.3.2   FPEC(AR-MODEL FITTING FOR CONTROL)
C       THIS PROGRAM PERFORMS FPEC(AR-MODEL FITTING FOR CONTROL)
C       COMPUTATION.
C       BESIDES THE OUTPUTS OF PROGRAM 5.1.2   MULCOR, THE FOLLOWING
C       INPUTS ARE REQUIRED:
C       L: UPPER LIMIT OF MODEL ORDER M (LESS THAN 30)
C       IR: NUMBER OF CONTROLLED VARIABLES
C       IL: NUMBER OF MANINPULATED VARIABLES, IL=0 FOR MFPE COMPUTATION
C       INW(I): INDICATOR; FIRST IR INDICATE THE CONTROLLED VARIABLES
C       AND THE REST THE MANIPULATE VARIABLES WITHIN THE IPO VARIABLES
C       IN THE OUTPUT OF PROGRAM 5.1.2   MULCOR.
C       THE OUTPUTS ARE THE PREDICTION ERROR COVARIANCE MATRIX OSD AND
C       THE SET OF COEFFICIENT MATRICES A AND B TO BE USED IN
C.      PROGRAM 5.5.1   OPTIMAL CONTROLLER DESIGN.
        IMPLICIT REAL*8(A-H,O-Z)
        DIMENSION R1(30,10,10),A1(30,10,10),B1(30,10,10),AO(30,10,10)
        DIMENSION SD(10,10),SE(10,10),SF(10,10),OSD(10,10)
        DIMENSION XSD(10,10),XSF(10,10),D(10,10),E(10,10),Z1(10,10)
        DIMENSION INW(10),C1(10,10)
C       ABSOLUTE DIMENSIONS USED FOR SUBROUTINE CALL
        MJ=10
        MJO=30
C       INITIAL CONDITION INPUT
        READ(5,1) L,IR,IL
        IP=IR+IL
        READ(5,1) (INW(I),I=1,IP)
C       READING THE OUTPUTS OF PROGRAM 5.1.2 MULCOR
        READ(5,1) N,LAGH,IPO
        L1=L+1
        LAGH1=LAGH+1
        CALL RECOVA(R1,LAGH1,L1,IPO,MJO,MJ)
        DO 10 II=1,L1
        DO 20 I=1,IPO
        DO 20 J=1,IPO
     20 C1(I,J)=R1(II,I,J)
C       MATRIX REARRANGEMENT BY INW
        CALL REARRA(C1,INW,IPO,IP,MJ)
        DO 21 I=1,IP
        DO 21 J=1,IP
     21 R1(II,I,J)=C1(I,J)
     10 CONTINUE
C       INITIAL CONDITION AND COVARIANCE PRINT OUT
        WRITE(6,39)
        WRITE(6,40)
        WRITE(6,41) N,L,IR,IL
        WRITE(6,259) (INW(I),I=1,IP)
        WRITE(6,42)
        CALL PRMAT3(R1,L1,IP,IP,1,MJO,MJ,MJ)
C       INITIAL SD, SF, SE COMPUTATION
        DO 330 II=1,IP
        DO 330 JJ=1,IP
        SD(II,JJ)=R1(1,II,JJ)
        SF(II,JJ)=SD(II,JJ)
        SE(II,JJ)=R1(2,II,JJ)
        XSD(II,JJ)=SD(II,JJ)
    330 XSF(II,JJ)=SF(II,JJ)
C       0-TH STEP COMPUTATION
        IFPEC=0
        MS=0
C       OFPEC, ORFPEC COMPUTATION
        CALL SFPEC(SD,N,IP,IR,MS,OFPEC,ORFPEC,OOFPEC,MJ)
C       OFPEC, ORFPEC PRINT OUT
        WRITE(6,600)
        WRITE(6,264) MS,OFPEC,ORFPEC
```

```
C      ITERATION M=1 TO L
       DO 400 M=1,L
C      INVERSE OF SD, SF COMPUTATION
       CALL INVDET(XSD,SDDET,IP,MJ)
       CALL INVDET(XSF,SFDET,IP,MJ)
C      D, E, SD, SF COMPUTATION
       CALL MULPLY(SE,XSF,D,IP,IP,IP,MJ,MJ,MJ)
       CALL TRAMDL(SE,XSD,E,IP,IP,IP,MJ,MJ,MJ)
       CALL TRAMDR(D,SE,Z1,IP,IP,IP,MJ,MJ,MJ)
       CALL SUBTAL(SD,Z1,IP,IP,MJ,MJ)
       CALL MULPLY(E,SE,Z1,IP,IP,IP,MJ,MJ,MJ)
       CALL SUBTAL(SF,Z1,IP,IP,MJ,MJ)
       MS=M
       DO 410 II=1,IP
       DO 410 JJ=1,IP
       XSD(II,JJ)=SD(II,JJ)
  410  XSF(II,JJ)=SF(II,JJ)
C      FPEC,RFPEC COMPUTATION
       CALL SFPEC(SD,N,IP,IR,MS,FPEC,RFPEC,OOFPEC,MJ)
C      FPEC,RFPEC PRINT OUT
       WRITE(6,264) MS,FPEC,RFPEC
C      FORWARD AND BACKWARD PREDICTOR COMPUTATION
       CALL COEFAB(A1,B1,D,E,MS,IP,MJ0,MJ)
C      MIN.FPEC, MIN.RFPEC COMPUTATION
       IF(OFPEC.LE.FPEC) GO TO 440
       OFPEC=FPEC
       ORFPEC=RFPEC
       IFPEC=M
       DO 560 II=1,IR
       DO 560 JJ=1,IR
  560  OSD(II,JJ)=SD(II,JJ)
       DO 561 I=1,M
       DO 562 II=1,IR
       DO 562 JJ=1,IP
  562  AO(I,II,JJ)=A1(I,II,JJ)
  561  CONTINUE
  440  IF(M.EQ.L) GO TO 400
C      SE COMPUTATION
       CALL NEWSE(A1,R1,SE,MS,IP,MJ0,MJ)
  400  CONTINUE
C      MIN.FPEC, MIN.RFPEC PRINT OUT
       WRITE(6,607) OFPEC,ORFPEC,IFPEC
C      OSD, AO PRINT AND PUNCH OUT
       WRITE(6,608)
       CALL SUBMPR(OSD,IR,IR,MJ,MJ)
  690  WRITE(7,1) N,IFPEC,IR,IL
       DO 680 II=1,IR
  680  WRITE(7,2) (OSD(II,JJ),JJ=1,IR)
       IF(IFPEC.LE.0) GO TO 699
       WRITE(6,609)
       CALL PRMAT3(AO,IFPEC,IR,IP,0,MJ0,MJ,MJ)
       DO 581 I=1,IFPEC
       DO 582 II=1,IR
  582  WRITE(7,2) (AO(I,II,JJ),JJ=1,IP)
  581  CONTINUE
  699  STOP
    1  FORMAT(10I5)
    2  FORMAT(4D20.10)
   39  FORMAT(1H1,50HPROGRAM 5.3.2     FPEC(AR-MODEL FITTING FOR CONTROL))
   40  FORMAT(1H0,17HINITIAL CONDITION)
   41  FORMAT(1H0,2HN=,I5,5X,2HL=,I5,5X,3HIR=,I5,5X,3HIL=,I5)
   42  FORMAT(//1H0,17HCOVARIANCE MATRIX)
  264  FORMAT(1H ,I5,2X,2D14.5)
  259  FORMAT(/1H0,6HINW(I),5X,10I5)
  600  FORMAT(////1H0,4X,1HI,12X,4HFPEC,9X,5HRFPEC)
```

```
  607 FORMAT(1H0,13HMINIMUM FPEC=,D12,5,2X,14HMINIMUM RFPEC=,D12,5,2X,14
     AHATTAINED AT M=,I5)
  608 FORMAT(//1H0,10X,10HOSD(II,JJ))
  609 FORMAT(//1H0,10X,10H(A(I)B(I)))
      END
C
      SUBROUTINE COEFAB(A1,B1,D,E,MS,K,MJ0,MJ)
C     THIS SUBROUTINE COMPUTES FORWARD(A) AND BACKWARD(B) PREDICTOR
C     COEFFICIENTS.
      IMPLICIT REAL*8(A-H,O-Z)
      DIMENSION A1(MJ0,MJ,MJ),B1(MJ0,MJ,MJ)
      DIMENSION D(MJ,MJ),E(MJ,MJ)
      DIMENSION A(10,10),B(10,10),Z1(10,10),Z2(10,10)
      IF(MS.EQ.1) GO TO 40
      MSM1=MS-1
      DO 10 I=1,MSM1
      MMI=MS-I
      DO 20 II=1,K
      DO 20 JJ=1,K
      A(II,JJ)=A1(I,II,JJ)
   20 B(II,JJ)=B1(MMI,II,JJ)
      CALL MULPLY(D,B,Z1,K,K,K,MJ,MJ,MJ)
      CALL MULPLY(E,A,Z2,K,K,K,MJ,MJ,MJ)
      CALL SUBTAL(A,Z1,K,K,MJ,MJ)
      CALL SUBTAL(B,Z2,K,K,MJ,MJ)
      DO 21 II=1,K
      DO 21 JJ=1,K
      A1(I,II,JJ)=A(II,JJ)
   21 B1(MMI,II,JJ)=B(II,JJ)
   10 CONTINUE
   40 DO 30 II=1,K
      DO 30 JJ=1,K
      A1(MS,II,JJ)=D(II,JJ)
   30 B1(MS,II,JJ)=E(II,JJ)
      RETURN
      END
```

```
      SUBROUTINE NEWSE(A1,R1,SE,MS,K,MJ0,MJ)
C     SE COMPUTATION
      IMPLICIT REAL*8(A-H,O-Z)
      DIMENSION A1(MJ0,MJ,MJ),R1(MJ0,MJ,MJ)
      DIMENSION SE(MJ,MJ)
      DIMENSION A(10,10),R(10,10),Z(10,10)
      DO 10 II=1,K
      DO 10 JJ=1,K
   10 Z(II,JJ)=0.0
      MSP2=MS+2
      DO 11 I=1,MS
      MMI=MSP2-I
      DO 12 II=1,K
      DO 12 JJ=1,K
      A(II,JJ)=A1(I,II,JJ)
   12 R(II,JJ)=R1(MMI,II,JJ)
      CALL MULPLY(A,R,SE,K,K,MJ,MJ,MJ)
   11 CALL MATADL(Z,SE,K,K,MJ,MJ)
      DO 14 II=1,K
      DO 14 JJ=1,K
   14 R(II,JJ)=R1(MSP2,II,JJ)
      CALL SUBTAC(R,Z,SE,K,K,MJ,MJ)
      RETURN
      END
C
      SUBROUTINE RECOVA(X,LAGH1,L1,IP0,MJ0,MJ)
C     COVARIANCE MATRIX INPUT
      IMPLICIT REAL*8(A-H,O-Z)
      DIMENSION X(MJ0,MJ,MJ)
      DIMENSION C(501),S(501)
      DO 10 II=1,IP0
C     AUTO COVARIANCE INPUT
      READ(5,1) IR0,IC0
      READ(5,2) (C(I),I=1,LAGH1)
      DO 20 I=1,L1
   20 X(I,II,II)=C(I)
      IF(II.EQ.1) GO TO 10
      IM1=II-1
      DO 11 JJ=1,IM1
C     CROSS COVARIANCE INPUT
      READ(5,1) IR1,IC1
      READ(5,2) (C(I),I=1,LAGH1)
      READ(5,1) IR2,IC2
      READ(5,2) (S(I),I=1,LAGH1)
      DO 30 I=1,L1
      X(I,II,JJ)=C(I)
   30 X(I,JJ,II)=S(I)
   11 CONTINUE
   10 CONTINUE
      RETURN
    1 FORMAT(10I5)
    2 FORMAT(4D20.10)
      END
```

```
      SUBROUTINE SFPEC(SD,N,K,IR,MS,Z,RZ,OOZ,MJ)
C     FPEC COMPUTATION
      IMPLICIT REAL*8(A-H,O-Z)
      DIMENSION SD(MJ,MJ)
      DIMENSION SD1(10,10)
      AN=N
      KM=K*MS
      ANP=N+1+KM
      ANM=N-1-KM
      AP=ANP/ANM
      APR=AP**IR
      DO 9 I=1,IR
      DO 9 J=1,IR
    9 SD1(I,J)=SD(I,J)
      CALL SUBDET(SD1,SDRM,IR,MJ)
      Z=APR*SDRM
      IF(MS.NE.0) GO TO 10
      OOZ=1.0/Z
   10 RZ=Z*OOZ
      RETURN
      END
C
      SUBROUTINE SUBDET(X,XDETMI,MM,MJ)
C     THIS SUBROUTINE COMPUTES THE DETERMINANT OF UPPER LEFT MM X MM
C     OF X.  FOR GENERAL USE STATEMENTS 20-21 SHOULD BE RESTORED.
C     X: ORIGINAL MATRIX
C     XDETMI: DETERMINANT OF UPPER LEFT MM X MM OF X
C     MJ: ABSOLUTE DIMENSION OF X IN THE MAIN ROUTINE
      IMPLICIT REAL*8(X)
      DIMENSION X(MJ,MJ)
      XDETMI=1.0
      IF(MM.EQ.1) GO TO 18
      MM1=MM-1
      DO 10 I=1,MM1
C  20 IF(X(I,I).NE.0.0) GO TO 11
C     DO 12 J=I,MM
C     IF(X(I,J).EQ.0.0) GO TO 12
C     JJ=J
C     GO TO 13
C  12 CONTINUE
C     XDETMI=0.0
C     GO TO 17
C  13 DO 14 K=I,MM
C     XXC=X(K,JJ)
C     X(K,JJ)=X(K,I)
C  14 X(K,I)=XXC
C  21 XDETMI=-XDETMI
   11 XDETMI=XDETMI*X(I,I)
      XC=1.0/X(I,I)
      I1=I+1
      DO 15 J=I1,MM
      XXC=X(J,I)*XC
      DO 16 K=I1,MM
   16 X(J,K)=X(J,K)-X(I,K)*XXC
   15 CONTINUE
   10 CONTINUE
   18 XDETMI=XDETMI*X(MM,MM)
   17 RETURN
      END
```

```
      SUBROUTINE SUBTAC(X,Y,Z,MM,NN,MJ1,MJ2)
C     MATRIX SUBTRACTION
C     Z=X-Y
C     (UPPER LEFT MM X NN OF Z)=(UPPER LEFT MM X NN OF X)-(UPPER LEFT
C     MM X NN OF Y).
C     (MJ1,MJ2): ABSOLUTE DIMENSION OF X, Y AND Z IN THE MAIN ROUTINE
      IMPLICIT REAL*8(A-H,O-Z)
      DIMENSION X(MJ1,MJ2),Y(MJ1,MJ2),Z(MJ1,MJ2)
      DO 10 I=1,MM
      DO 10 J=1,NN
   10 Z(I,J)=X(I,J)-Y(I,J)
      RETURN
      END
C
      SUBROUTINE SUBTAL(X,Y,MM,NN,MJ1,MJ2)
C     MATRIX SUBTRACTION
C     X=X-Y
C     (UPPER LEFT MM X NN OF X)=(UPPER LEFT MM X NN OF X)-(UPPER LEFT
C     MM X NN OF Y).
C     (MJ1,MJ2): ABSOLUTE DIMENSION OF X AND Y IN THE MAIN ROUTINE
      IMPLICIT REAL*8(A-H,O-Z)
      DIMENSION X(MJ1,MJ2),Y(MJ1,MJ'
      DO 10 I=1,MM
      DO 10 J=1,NN
   10 X(I,J)=X(I,J)-Y(I,J)
      RETURN
      END
C
      SUBROUTINE TRAMDR(X,Y,Z,MM,NN,NC,MJ1,MJ2,MJ3)
C     TRANSPOSE MULTIPLY (RIGHT)
C     Z=X*Y'
C     (UPPER LEFT MM X NC OF Z)=(UPPER LEFT MM X NN OF X)*(UPPER LEFT
C     NC X NN OF Y)',
C     (MJ1,MJ2): ABSOLUTE DIMENSION OF X IN THE MAIN ROUTINE
C     (MJ3,MJ2): ABSOLUTE DIMENSION OF Y IN THE MAIN ROUTINE
C     (MJ1,MJ3): ABSOLUTE DIMENSION OF Z IN THE MAIN ROUTINE
      IMPLICIT REAL*8(A-H,O-Z)
      DIMENSION X(MJ1,MJ2),Y(MJ3,MJ2),Z(MJ1,MJ3)
      DO 10 I=1,MM
      DO 11 J=1,NC
      SUM=0.0
      DO 12 K=1,NN
   12 SUM=SUM+X(I,K)*Y(J,K)
      Z(I,J)=SUM
   11 CONTINUE
   10 CONTINUE
      RETURN
      END
```

```
      PROGRAM MULNOS
C     PROGRAM 5.3.3   MULTIPLE UNOISE
C     THIS PROGRAM COMPUTES RELATIVE POWER CONTRIBUTIONS IN DIFFERENTIAL
C     AND INTEGRATED FORM, ASSUMING THE ORTHOGONALITY BETWEEN NOISE
C     SOURCES.
C     THE PROGRAM OPERATES ON THE OUTPUT OF PROGRAM 5.3.2 FPEC WITH
C     IL=0.
C     THE RESULTS ARE GIVEN AT FREQUIENCIES I/(2*H).
      IMPLICIT REAL*8(A-H,O-W)
      IMPLICIT COMPLEX*16(X-Z)
      INTEGER H,H1
      COMMON G,LG,GR,GI,H,JJF
      DIMENSION SD(10,10),A(30,10,10),X(10,10)
      DIMENSION G(31),RS(10,10),R(10,10)
C     ABSOLUTE DIMENSIONS USED FOR SUBROUTINE CALL
      MJ0=30
      MJ1=10
C     H SPECIFICATION
      READ(5,1) H
C     READING THE OUTPUT OF PROGRAM 5.3.2 FPEC WITH IL=0
      READ(5,1) N,L,IP
C     SD INPUT
      CALL REMATX(SD,IP,IP,1,MJ1,MJ1)
C     A INPUT
      CALL REMAT3(A,L,IP,IP,1,MJ0,MJ1,MJ1)
  310 H1=H+1
C     SD NORMALIZATION
      DO 100 I=1,IP
      DO 100 J=1,IP
  100 RS(I,J)=SD(I,J)/DSQRT(SD(I,I)*SD(J,J))
C     INITIAL CONDITION PRINT OUT
      WRITE(6,59)
      WRITE(6,60)
      WRITE(6,61) H,N,L,IP
      WRITE(6,161)
      CALL SUBMPR(SD,IP,IP,MJ1,MJ1)
C     NORMALIZED SD PRINT OUT
      WRITE(6,162)
      CALL SUBMPR(RS,IP,IP,MJ1,MJ1)
C     A PRINT OUT
      WRITE(6,420)
      CALL PRMAT3(A,L,IP,IP,0,MJ0,MJ1,MJ1)
  410 DO 10 JF=1,H1
      JJF=JF
C     AF COMPUTATION
      DO 40 II=1,IP
      DO 41 JJ=1,IP
      IF(II.NE.JJ) GO TO 42
      G(1)=1.0
      GO TO 43
   42 G(1)=0.0
   43 DO 45 I=1,L
      I1=I+1
   45 G(I1)=-A(I,II,JJ)
   44 LG=L
      CALL FGER1
      X(II,JJ)=DCMPLX(GR,GI)
   41 CONTINUE
   40 CONTINUE
```

```
C     INVERSE OF AF (COMPLEX) COMPUTATION
      CALL INVDET(X,XDET,IP,MJ1)
C     RELATIVE POWER CONTRIBUTIONS COMPUTATION
      CALL SUBNOS(X,SD,IP,RS,R,MJ1)
C     RELATIVE POWER CONTRIBUTIONS PRINT OUT
      JFM1=JF-1
      WRITE(6,65) JFM1
      WRITE(6,165)
      CALL SUBMPR(RS,IP,IP,10,10)
      WRITE(6,166)
      CALL SUBMPR(R,IP,IP,10,10)
   10 CONTINUE
      STOP
    1 FORMAT(10I5)
    2 FORMAT(4D20.10)
   59 FORMAT(1H1,31HPROGRAM 5.3.3    MULTIPLE UNOISE)
   60 FORMAT(1H0,17HINITIAL CONDITION)
   61 FORMAT(1H0,2HH=,I5,5X,2HN=,I5,5X,2HL=,I5,5X,3HIP=,I5)
  161 FORMAT(/1H0,7HSD(I,J))
  162 FORMAT(/1H0,13HNORMALIZED SD)
  420 FORMAT(/1H0,6HA(I,J))
   65 FORMAT(///1H0,2HF=,I5)
  165 FORMAT(/1H0,40HDIFFERENTIAL RELATIVE POWER CONTRIBUTION)
  166 FORMAT(/1H0,38HINTEGRATED RELATIVE POWER CONTRIBUTION)
      END
C
      SUBROUTINE SUBNOS(X,SD,IP,RS,R,MJ)
C     THIS SUBROUTINE COMPUTES RELATIVE POWER CONTRIBUTIONS.
C     MJ: ABSOLUTE DIMENSION OF X IN THE MAIN ROUTINE
C     IP: DIMENSION OF RS1 OR RL (LESS THAN 11)
      IMPLICIT REAL*8(A-H,O-W)
      IMPLICIT COMPLEX*16(X-Z)
      DIMENSION X(MJ,MJ)
      DIMENSION SD(MJ,MJ),RS(MJ,MJ),R(MJ,MJ)
      DIMENSION RS1(10),RL(10)
      DO 10 II=1,IP
      SUM=0.0
      DO 11 JJ=1,IP
      RX=DREAL(X(II,JJ))
      RIX=DIMAG(X(II,JJ))
      RS1(JJ)=(RX**2+RIX**2)*SD(JJ,JJ)
      SUM=SUM+RS1(JJ)
   11 RL(JJ)=SUM
      RCONST=1.0/RL(IP)
      DO 14 JJ=1,IP
   14 RS(II,JJ)=RS1(JJ)*RCONST
      DO 12 LL=1,IP
   12 R(II,LL)=RL(LL)*RCONST
   10 CONTINUE
      RETURN
      END
```

```
      PROGRAM DECONV
C     PROGRAM 5.3.4    DECONVOLUTION
C     THIS PROGRAM COMPUTES IMPULSE RESPONSES OF FEEDBACK LOOPS USING
C     THE OUTPUT OF PROGRAM 5.3.2 FPEC (WITH IL=0).
C     ONE CARD WITH INFORMATION OF ML, MAXIMUM LENGTH OF RESPONSE,
C     SHOULD BE ADDED ON TOP OF THE OUTPUT OF PROGRAM FPEC (WITH IL=0)
C     TO FORM THE INPUT TO THIS PROGRAM.
C     THE OUTPUTS ARE SAIJ(M), THE RESPONSE OF XI TO XJ, M=1 TO ML AND
C     I,J=1 TO IP.
      IMPLICIT REAL*8(A-H,O-Z)
      DIMENSION A1(30,10,10),A(30),C(30),SA(101),C1(101)
      DIMENSION OSD(10,10)
C     ABSOLUTE DIMENSIONS USED FOR SUBROUTINE CALL
      MJO=30
      MJ1=10
C     INITIAL CONDITION INPUT
      READ(5,1) ML
C     READING THE OUTPUTS OF PROGRAM 5.3.2 FPEC (WITH IL=0)
      READ(5,1) N,MO,IP
      CALL REMATX(OSD,IP,IP,1,MJ1,MJ1)
C     ABOVE INPUT IS NONEFFECTIVE.
      CALL REMAT3(A1,MO,IP,IP,1,MJO,MJ1,MJ1)
C     INITIAL CONDITION PRINT OUT
      WRITE(6,60)
      WRITE(6,61)
      WRITE(6,62) ML
      WRITE(6,63) N,MO,IP
      WRITE(6,64)
      CALL PRMAT3(A1,MO,IP,IP,0,MJO,MJ1,MJ1)
C     COMPUTATION START
      MLP1=ML+1
      DO 10 II=1,IP
C     C ARRANGEMENT
      DO 11 L=1,MO
   11 C(L)=A1(L,II,II)
      DO 20 JJ=1,IP
      IF(II.EQ.JJ) GO TO 20
C     A ARRANGEMENT
      DO 22 L=1,MO
   22 A(L)=A1(L,II,JJ)
C     INITIAL CONDITIONING
      DO 30 I=1,MLP1
      C1(I)=0.0
   30 SA(I)=0.0
C     SA COMPUTATION
      SA(2)=A(1)
      DO 40 I=2,ML
      IM1=I-1
      IF(IM1.GT.MO) GO TO 42
      C1(I)=C(IM1)
   42 IP1=I+1
      CALL CONVOL(C1,SA,SA2,IP1)
      IF(I.LE.MO) SA2=A(I)+SA2
   40 SA(IP1)=SA2
```

```
C      SA PRINT OUT
       WRITE(6,160) II,JJ
       WRITE(6,162)
       CALL PRCOL1(SA,2,MLP1,1)
   20 CONTINUE
   10 CONTINUE
       STOP
    1 FORMAT(10I5)
   60 FORMAT(1H1,29HPROGRAM 5.3.4    DECONVOLUTION)
   61 FORMAT(1H0,17HINITIAL CONDITION)
   62 FORMAT(1H0,3HML=,I5)
   63 FORMAT(1H ,2HN=,I5,5X,3HM0=,I5,5X,3HIP=,I5)
   64 FORMAT(1H0,7HA1(I,J))
  160 FORMAT(//1H0,2HI=,I5,5X,2HJ=,I5)
  162 FORMAT(1H0,4X,1HM,9X,5HSA(M))
       END
C
       SUBROUTINE CONVOL(A,B,SUM,K1)
C      THIS SUBROUTINE COMPUTES CONVOLUTION.
C      C(K)=A(0)B(K)+A(1)B(K-1)+...+A(K-1)B(1)+A(K)B(0)
C      K1: K PLUS 1
       IMPLICIT REAL*8(A-H,O-Z)
       DIMENSION A(K1),B(K1)
       K2=K1+1
       SUM=0.0
       DO 10 I=1,K1
       KI=K2-I
   10 SUM=SUM+A(I)*B(KI)
       RETURN
       END
```

5.4   Computation of Spectra through Autoregression

The spectrum is determined when the coefficients of the autoregressive (AR) model and the variance (covariance matrix) of the white noise are given. Here we give a program for the spectral density calculation of an autoregressive moving average (AR-MA) model which is obtained by replacing the white noise term of the autoregressive model by the moving average. This gives the so called rational spectrum. When the order of the moving average part of the AR-MA model is set equal to zero, the model reduces to an autoregressive model. Similarly, if the order of autoregression is set equal to zero, a moving average (MA) model is obtained.

The estimate of the cross-spectral density obtained by this program can be used as the input to the frequency response computation of 5.2. However, the output of the error evaluation of the estimate becomes invalid. The Goertzel method of Fourier transformation is adopted in this program.

[Program]
    5.4.1 RASPEC Rational Spectrum Computation (uni-variate)
    5.4.2 MULRSP Rational Spectrum Computation (multi-variate)

```
      PROGRAM RASPEC
C     PROGRAM 5.4.1    RATIONAL SPECTRUM
C     THIS PROGRAM COMPUTES POWER SPECTRUM OF AR-MA PROCESS
C     X(N)=A(1)X(N-1)+...+A(L)X(N-L)+E(N)+B(1)E(N-1)+...+B(K)E(N-K)
C     WHERE E(N) IS A WHITE NOISE WITH ZERO MEAN AND VARIANCE EQUAL TO
C     SGME2.   OUTPUTS PXX(I) ARE GIVEN AT FREQUENCIES I/(2*H)
C     I=0,1,...,H.
C     REQUIRED INPUTS ARE:
C     L,K,H,SGME2,(A(I),I=1,L), AND (B(I),I=1,K).
C     0 IS ALLOWABLE AS L AND/OR K.
      IMPLICIT REAL*8(A-H,O-Z)
      INTEGER H,H1
      DIMENSION A(501),B(501)
      DIMENSION G(501),GR1(501),GI1(501),GR2(501),GI2(501)
      DIMENSION PXX(510)
C     H SPECIFICATION
      READ(5,1) H
C     SGME2 AND A INPUT
C     THE OUTPUTS OF PROGRAM 5.3.1 FPE AUTO CAN BE USED AS THE FOLLOWING
C     INPUTS WITH K=0.
      READ(5,1) N,L
      READ(5,2) SGME2
      IF(L.LE.0) GO TO 300
      READ(5,2) (A(I),I=1,L)
C     K INPUT
  300 READ(5,1) K
      IF(K.LE.0) GO TO 310
      READ(5,2) (B(I),I=1,K)
  310 H1=H+1
      L1=L+1
      K1=K+1
      G(1)=1.0
      IF(L.LE.0) GO TO 400
      DO 10 I=1,L
      I1=I+1
   10 G(I1)=-A(I)
  400 CALL FOUGER(G,L1,GR1,GI1,H1)
      G(1)=1.0
      IF(K.LE.0) GO TO 410
      DO 20 I=1,K
      I1=I+1
   20 G(I1)=B(I)
  410 CALL FOUGER(G,K1,GR2,GI2,H1)
      DO 30 I=1,H1
   30 PXX(I)=(GR2(I)**2+GI2(I)**2)/(GR1(I)**2+GI1(I)**2)*SGME2
      WRITE(6,60)
      WRITE(6,160)
      WRITE(6,61) L,K,H
      WRITE(6,164) SGME2
      IF(L.LE.0) GO TO 500
      WRITE(6,62)
      CALL PRCOL1(A,1,L,0)
  500 IF(K.LE.0) GO TO 510
      WRITE(6,63)
      CALL PRCOL1(B,1,K,0)
  510 WRITE(6,64)
      CALL PRCOL1(PXX,1,H1,1)
      STOP
```

```
    1 FORMAT(10I5)
    2 FORMAT(4D20.10)
   60 FORMAT(1H1,33HPROGRAM 5.4.1    RATIONAL SPECTRUM)
   61 FORMAT(1H0,2HL=,I5,2X,2HK=,I5,2X,2HH=,I5)
   62 FORMAT(1H0,4X,1HI,12X,4HA(I))
   63 FORMAT(1H0,4X,1HI,12X,4HB(I))
   64 FORMAT(1H0,4X,1HI,10X,6HPXX(I))
  160 FORMAT(1H0,17HINITIAL CONDITION)
  164 FORMAT(1H0,6HSGME2=,D12.5)
      END
C
      SUBROUTINE FOUGER(G,LGP1,FC,FS,LF1)
C     FOURIER TRANSFORM (GOERTZEL METHOD)
C     THIS SUBROUTINE COMPUTES FOURIER TRANSFORM OF G(I),I=0,1,...,LG AT
C     FREQUENCIES K/(2*LF),K=0,1,...,LF AND RETURNS COSIN TRANSFORM IN
C     FC(K) AND SIN TRANSFORM IN FS(K).
      IMPLICIT REAL*8(A-H,O-Z)
      DIMENSION G(LGP1),FC(LF1),FS(LF1)
      LG=LGP1-1
      LF=LF1-1
C     REVERSAL OF G(I),I=1,...,LGP1 INTO G(LG3-1)    LG3=LGP1+1
      IF(LGP1.LE.1) GO TO 110
      LG3=LGP1+1
      LG4=LGP1/2
      DO 100 I=1,LG4
      I2=LG3-I
      T=G(I)
      G(I)=G(I2)
  100 G(I2)=T
  110 PI=3.1415926536
      ALF=LF
      T=PI/ALF
      DO 10 K=1,LF1
      AK=K-1
      TK=T*AK
      CK=DCOS(TK)
      SK=DSIN(TK)
      CK2=CK+CK
      UM2=0.0
      UM1=0.0
      IF(LG.EQ.0) GO TO 12
      DO 11 I=1,LG
      UM0=CK2*UM1-UM2+G(I)
      UM2=UM1
   11 UM1=UM0
   12 FC(K)=CK*UM1-UM2+G(LGP1)
      FS(K)=SK*UM1
   10 CONTINUE
      RETURN
      END
```

```
      PROGRAM MULRSP
C     PROGRAM 5.4.2    MULTIPLE RATIONAL SPECTRUM
C     THIS PROGRAM COMPUTES RATIONAL SPECTRUM FOR IP-DIMENSIONAL
C     AR-MA PROCESS
C     X(N)=A(1)X(N-1)+...+A(L)X(N-L)+E(N)+B(1)E(N-1)+...+B(K)E(N-K),
C     WHERE E(N) IS A WHITE NOISE WITH ZERO MEAN VECTOR AND COVARIANCE
C     MATRIX SD.
C     OUTPUTS ARE SPECTRUM MATRIX P(I) AT FREQUENCIES I/(2*H)
C     (I=0,1,...,H).
      IMPLICIT REAL*8(A-H,O-W)
      IMPLICIT COMPLEX*16(X-Z)
      INTEGER H,H1
      COMMON G,LG,GR,GI,H,JJF
      DIMENSION SD(10,10),A(30,10,10),B(30,10,10),X(10,10),Y(10,10)
      DIMENSION Z(10,10),G(31),CH(10,10)
C     ABSOLUTE DIMENSIONS USED FOR SUBROUTINE CALL
      MJ0=30
      MJ1=10
C     H SPECIFICATION
      READ(5,1) H
C     SD AND A INPUT
C     THE OUTPUTS OF PROGRAM 5.3.2 FPEC(WITH IL=0) CAN BE USED AS THE
C     FOLLWOING INPUTS WITH K=0.
      READ(5,1) N,L,IP
C     SD INPUT
      CALL REMATX(SD,IP,IP,1,MJ1,MJ1)
      IF(L.LE.0) GO TO 300
C     A INPUT
      CALL REMAT3(A,L,IP,IP,1,MJ0,MJ1,MJ1)
C     K INPUT
  300 READ(5,1) K
      IF(K.LE.0) GO TO 310
C     B INPUT
      CALL REMAT3(B,K,IP,IP,1,MJ0,MJ1,MJ1)
  310 H1=H+1
C     INITIAL CONDITION PRINT OUT
      WRITE(6,59)
      WRITE(6,60)
      WRITE(6,61) H,N,L,IP,K
      WRITE(6,161)
      CALL SUBMPR(SD,IP,IP,MJ1,MJ1)
      IF(L.LE.0) GO TO 400
C     A PRINT OUT
      WRITE(6,420)
      CALL PRMAT3(A,L,IP,IP,0,MJ0,MJ1,MJ1)
  400 IF(K.LE.0) GO TO 410
C     B PRINT OUT
      WRITE(6,430)
      CALL PRMAT3(B,K,IP,IP,0,MJ0,MJ1,MJ1)
C     SPECTRUM COMPUTATION
  410 DO 10 JF=1,H1
      JJF=JF
C     SD STORE
      DO 631 II=1,IP
      DO 631 JJ=1,IP
  631 Y(II,JJ)=SD(II,JJ)
      IF(K.GT.0) GO TO 100
      DO 110 II=1,IP
      DO 110 JJ=1,IP
  110 Z(II,JJ)=SD(II,JJ)
      GO TO 224
```

```
C     BF CONPUTATION
  100 DO 20 II=1,IP
      DO 21 JJ=1,IP
      IF(II.NE.JJ) GO TO 22
      G(1)=1.0
      GO TO 23
   22 G(1)=0.0
   23 DO 25 I=1,K
      I1=I+1
   25 G(I1)=B(I,II,JJ)
   24 LG=K
      CALL FGER1
      X(II,JJ)=DCMPLX(GR,GI)
   21 CONTINUE
   20 CONTINUE
C     BF*SD*CONJG(BF') COMPUTATION
      CALL XYCTRX(X,Y,Z,IP,IP,MJ1,MJ1)
  224 IF(L.GT.0) GO TO 120
      DO 130 II=1,IP
      DO 130 JJ=1,IP
  130 Y(II,JJ)=Z(II,JJ)
      GO TO 244
C     AF COMPUTATION
  120 DO 40 II=1,IP
      DO 41 JJ=1,IP
      IF(II.NE.JJ) GO TO 42
      G(1)=1.0
      GO TO 43
   42 G(1)=0.0
   43 DO 45 I=1,L
      I1=I+1
   45 G(I1)=-A(I,II,JJ)
   44 LG=L
      CALL FGER1
      X(II,JJ)=DCMPLX(GR,GI)
   41 CONTINUE
   40 CONTINUE
C     INVERSE OF AF (COMPLEX) COMPUTATION
      CALL INVDET(X,XDET,IP,MJ1)
C     (INVERSE OF AF)*(BF*SD*CONJG(BF'))*CONJG((INVERSE OF AF)')
C     COMPUTATION
      CALL XYCTRX(X,Z,Y,IP,IP,MJ1,MJ1)
C     SIMPLE COHERENCE COMPUTATION
  244 CH(1,1)=1.0
      IF(IP.EQ.1) GO TO 260
      DO 50 II=2,IP
      IM1=II-1
      RYI=DREAL(Y(II,II))
      DO 51 JJ=1,IM1
      RYJ=DREAL(Y(JJ,JJ))
      RRYIJ=DREAL(Y(II,JJ))
      RIYIJ=DIMAG(Y(II,JJ))
      CH(II,JJ)=(RRYIJ**2+RIYIJ**2)/(RYI*RYJ)
   51 CH(JJ,II)=CH(II,JJ)
   50 CH(II,II)=1.0
```

```
C     RATIONAL SPECTRUM AND SIMPLE COHERENCE PRINT OUT
  260 JFM1=JF-1
      WRITE(6,65) JFM1
      WRITE(6,66)
      CALL PRCPMA(Y,IP,IP,MJ1,MJ1)
      WRITE(6,67)
      CALL SUPMPR(CH,IP,IP,MJ1,MJ1)
C
   10 CONTINUE
      STOP
    1 FORMAT(10I5)
   59 FORMAT(1H1,42HPROGRAM 5.4.2    MULTIPLE RATIONAL SPECTRUM)
   60 FORMAT(1H0,17HINITIAL CONDITION)
   61 FORMAT(1H0,2HH=,I5,5X,2HN=,I5,5X,2HL=,I5,5X,3HIP=,I5,5X,2HK=,I5)
   62 FORMAT(1H0,2HI=,I5)
   65 FORMAT(///1H0,2HF=,I5)
   66 FORMAT(1H0,5X,17HRATIONAL SPECTRUM)
   67 FORMAT(/1H0,5X,16HSIMPLE COHERENCE)
  161 FORMAT(//1H0,7HSD(I,J))
  420 FORMAT(//1H0,6HA(I,J))
  430 FORMAT(//1H0,6HB(I,J))
      END
C
      SUBROUTINE XYCTRX(X,Y,Z,MM,NN,MJ1,MJ2)
C     Z=X*Y*CONJG(X')
C     Y,Z: HERMITIAN
C     (UPPER LEFT MM X MM OF Z)=(UPPER LEFT MM X NN OF X)*(UPPER LEFT
C     NN X NN OF Y)*CONJG((UPPER LEFT MM X NN OF X)')
C     (MJ1,MJ2): ABSOLUTE DIMENSION OF X IN THE MAIN ROUTINE
C     (MJ2,MJ2): ABSOLUTE DIMENSION OF Y IN THE MAIN ROUTINE
C     (MJ1,MJ1): ABSOLUTE DIMENSION OF Z IN THE MAIN ROUTINE
C     MM,NN: SHOULD BE LESS THAN 11.
      IMPLICIT COMPLEX*16(X-Z)
      DIMENSION X(MJ1,MJ2),Y(MJ2,MJ2),Z(MJ1,MJ1)
      DIMENSION Y1(10,10)
      DO 10 I=1,MM
      DO 10 J=1,NN
      XSUM=0.0
      DO 12 K=1,NN
   12 XSUM=XSUM+X(I,K)*Y(K,J)
   10 Y1(I,J)=XSUM
      DO 110 I=1,MM
      DO 110 J=1,I
      XSUM=0.0
      DO 112 K=1,NN
  112 XSUM=XSUM+Y1(I,K)*DCONJG(X(J,K))
      Z(I,J)=XSUM
  110 Z(J,I)=DCONJG(Z(I,J))
      RETURN
      END
```

5.5 Optimal Controller Design

By adding the two positive definite matrices W and R that define the performance criterion and the length I of the basic time span for the design of the control to the output of the FPEC calculation by Program 5.3.2, the optimal feedback gain G is obtained by executing the calculation defined by the equations (23)–(27) of 3.5.3. This can be done by Program OPTDES. Program OPTSIM performs the simulation of the feedback system given by equation (1) of 3.5.4. Input representing the noise sources $\{W(s): s=1, 2, ..., N\}$ is required for this program. When a vector of random numbers simulating the noise source of the system is required for $W(s)$, refer to the method of generating a realization of $U(s)$ discussed in 3.2.3. Program WNOISE performs the necessary calculation.

[Programs]

    5.5.1 OPTDES Optimal Controller Design

    5.5.2 OPTSIM Optimal Control Simulation

    5.5.3 WNOISE White Noise Simulation

For the generation of uniform random numbers in WNOISE use is made of the program by Wingersky, discussed in the book by P. R. Lohnes and W. W. Cooley [3]. However, as is discussed by these authors, the generation of random numbers is a tricky subject and the reader must be careful in its use. It is advisable to check the validity of the simulation by observing the result of application of the random numbers to similar examples with theoretically confirmed structures.

If in Program 5.5.2 OPTSIM the initial values are put equal to the actual data of the past instead of O (zero vector), and the noise $W(s)$ is put equal to O, the output series provides a forecast of the future behavior of the feedback system. In this case, if the gain G is put equal to O (zero matrix), a forecast of the behavior of the system without feedback is obtained.

```
      PROGRAM OPTDES
C     PROGRAM 5.5.1    OPTIMAL CONTROLLER DESIGN
C     THIS PROGRAM COMPUTES OPTIMAL CONTROLLER GAIN MATRIX FOR
C     A QUADRATIC CRITERION DEFINED BY TWO POSITIVE DEFINITE MATRICES
C     Q1 AND R.
C     THE OUTPUT OF FPEC COMPUTATION IS USED.
C     THE FIRST IP VARIABLES SHOULD BE CONTROLLED VARIABLES.
C     IR: NUMBER OF CONTROLLED VARIABLES
C     L: NUMBER OF MANIPULATED VARIABLES
C     NS: NUMBER OF D.P. STAGES
C     N: LENGTH OF ORIGINAL DATA
C     M: ORDER OF THE MODEL WHICH GIVES THE MINIMUM OF FPEC
C     MATRIX P: MI OR P
C     GI: GAIN
      IMPLICIT REAL*8(A-H,O-Z)
      DIMENSION A(75,5),B(75,5),Q1(5,5),R(5,5)
      DIMENSION GI(5,75),GIT(75,5)
      DIMENSION GL(5,5),G3(5,75),GR(5,5),GLR(5,5)
      DIMENSION P(75,75),D(75)
C     ABSOLUTE DIMENSIONS USED FOR SUBROUTINE CALL
      MJ1=5
      MJ2=5
      MJ3=75
C     INITIAL CONDITION INPUT AND OUTPUT
      READ(5,1) IR,L,NS
      CALL REMTSB(Q1,IR,1,MJ1)
      CALL REMTSB(R,L,1,MJ2)
C     READING THE OUTPUTS OF PROGRAM 5.3.2 FPEC
      READ(5,1) N,M
      CALL REMATX(GR,IR,IR,1,MJ1,MJ1)
C     ABOVE INPUT IS NONEFFECTIVE.
      IM=0
      DO 8 JJ=1,M
      DO 9 I=1,IR
      II=IM+I
    9 READ(5,2) (A(II,J),J=1,IR),(B(II,J),J=1,L)
    8 IM=IM+IR
      MR=M*IR
      WRITE(6,60)
      WRITE(6,61)
      WRITE(6,362) N,M
      WRITE(6,62) IR,L,NS
      WRITE(6,65)
      CALL SUBMPR(Q1,IR,IR,MJ1,MJ1)
      WRITE(6,66)
      CALL SUBMPR(R,L,L,MJ2,MJ2)
      WRITE(6,63)
      CALL SUBMPR(A,MR,IR,MJ3,MJ1)
      WRITE(6,64)
      CALL SUBMPR(B,MR,L,MJ3,MJ2)
C     CONSTANTS FOR COMPUTATION
      MR1=MR-IR
      MM1=M-1
      MM2=M-2
      IBA=MM2*IR
C     INITIAL P COMPUTATION
      DO 130 I=1,MR
      DO 130 J=1,MR
  130 P(I,J)=0.0
      DO 131 I=1,IR
      DO 131 J=1,IR
  131 P(I,J)=Q1(I,J)
```

```
C     P ITERATION INS=1 TO NS
      DO 10 INS=1,NS
C     GI=(TRANSPOSE OF B)*P COMPUTATION
      CALL TRAMDL(B,P,GI,MR,L,MR,MJ3,MJ2,MJ3)
C     GL=GI*B COMPUTATION
      CALL MULTRB(GI,B,GL,L,MR,MJ2,MJ3)
      DO 20 I=1,L
      DO 20 J=1,I
   20 GL(J,I)=GL(I,J)
C     GL=GL+R COMPUTATION
      CALL MATADL(GL,R,L,L,MJ2,MJ2)
C     GL=(INVERSE OF GL) COMPUTATION
      CALL INVDET(GL,XDET,L,MJ2)
C     G3=GL*GI COMPUTATION
      CALL MULPLY(GL,GI,G3,L,MR,MJ2,MJ2,MJ3)
C     D= DIAGONAL OF P
      DO 30 I=1,MR
   30 D(I)=P(I,I)
C     INTERMEDIATE MATRIX MI COMPUTATION
C     P=(TRANSPOSE OF GI)*G3 (LOWER TRIANGLE) COMPUTATION
      CALL MULTRL(GI,G3,P,L,MR,MJ2,MJ3)
      DO 40 I=1,MR
   40 P(I,I)=D(I)-P(I,I)
      IF(MR.EQ.1) GO TO 260
      DO 41 I=2,MR
      IM1=I-1
      DO 41 J=1,IM1
      P(I,J)=P(J,I)-P(I,J)
   41 P(J,I)=P(I,J)
C     GIT=MI*A COMPUTATION
  260 CALL MULPLY(P,A,GIT,MR,MR,IR,MJ3,MJ3,MJ1)
C     GR=(TRANSPOSE OF A)*GIT COMPUTATION
      CALL MULTRL(A,GIT,GR,MR,IR,MJ3,MJ1)
      DO 150 I=1,IR
      DO 150 J=1,I
  150 GR(J,I)=GR(I,J)
C     GR=GR+Q1 COMPUTATION
      CALL MATADL(GR,Q1,IR,IR,MJ1,MJ1)
C     NEW P ARRANGEMENT
      IF(M.EQ.1) GO TO 261
      IIB=IBA
      DO 50 II=1,MM1
      IIO=M-II
      JJC=IIB
      DO 51 JJ=1,IIO
      DO 52 I=1,IR
      I1=IIB+I
      I2=I1+IR
      DO 52 J=1,IR
      J1=JJC+J
      J2=J1+IR
      P(I2,J2)=P(I1,J1)
   52 P(J2,I2)=P(I2,J2)
   51 JJC=JJC-IR
   50 IIB=IIB-IR
      DO 57 I=1,MR1
      I1=I+IR
      DO 57 J=1,IR
      P(I1,J)=GIT(I,J)
   57 P(J,I1)=P(I1,J)
  261 DO 56 I=1,IR
      DO 56 J=1,I
      P(I,J)=GR(I,J)
   56 P(J,I)=P(I,J)
   10 CONTINUE
```

```
C     GAIN COMPUTATION
C     GLR=-G3*A
      CALL MULPLY(G3,A,GLR,L,MR,IR,MJ2,MJ3,MJ1)
      DO 110 I=1,L
      DO 110 J=1,IR
  110 GI(I,J)=-GLR(I,J)
      IF(M.EQ.1) GO TO 262
      DO 111 I=1,L
      DO 111 J=1,MR1
      J1=IR+J
  111 GI(I,J1)=-G3(I,J)
C     GAIN PRINT AND PUNCH OUT
  262 WRITE(6,67)
      CALL SUBMPR(GI,L,MR,MJ2,MJ3)
      WRITE(7,1) N,M,IR,L,NS
      DO 232 I=1,IR
  232 WRITE(7,2) (Q1(I,J),J=1,IR)
      DO 233 I=1,L
  233 WRITE(7,2) (R(I,J),J=1,L)
      DO 230 I=1,MR
  230 WRITE(7,2) (A(I,J),J=1,IR)
      DO 231 I=1,MR
  231 WRITE(7,2) (B(I,J),J=1,L)
      DO 120 I=1,L
  120 WRITE(7,2) (GI(I,J),J=1,MR)
      STOP
    1 FORMAT(10I5)
    2 FORMAT(4D20.10)
   60 FORMAT(1H1,41HPROGRAM 5.5.1    OPTIMAL CONTROLLER DESIGN)
   61 FORMAT(1H0,17HINITIAL CONDITION)
   62 FORMAT(1H0,3HIR=,I5,5X,2HL=,I5,5X,3HNS=,I5)
  362 FORMAT(1H0,2HN=,I5,5X,2HM=,I5)
   63 FORMAT(//1H0,44HFIRST IR COLUMNS OF TRANSITION MATRIX (AI'S))
   64 FORMAT(//1H0,19HGAMMA MATRIX (BI'S))
   65 FORMAT(//1H0,10X,7HQ1(I,J))
   66 FORMAT(//1H0,10X,6HR(I,J))
   67 FORMAT(////1H0,13HGAIN MATRIX G)
      END
```

```
      SUBROUTINE MULTRB(X,Y,Z,MM,NN,MJ1,MJ2)
C     Z=X*Y
C     Z: SYMMETRIC
C     (LOWER TRIANGLE OF UPPER LEFT MM X MM OF Z)=(UPPER LEFT MM X NN OF
C     X)*(UPPER LEFT NN X MM OF Y)
C     (MJ1,MJ2): ABSOLUTE DIMENSION OF X IN THE MAIN ROUTINE
C     (MJ2,MJ1): ABSOLUTE DIMENSION OF Y IN THE MAIN ROUTINE
C     (MJ1,MJ1): ABSOLUTE DIMENSION OF Z IN THE MAIN ROUTINE
      IMPLICIT REAL*8(A-H,O-Z)
      DIMENSION X(MJ1,MJ2),Y(MJ2,MJ1),Z(MJ1,MJ1)
      DO 10 I=1,MM
      DO 10 J=1,I
      SUM=0.0
      DO 11 K=1,NN
   11 SUM=SUM+X(I,K)*Y(K,J)
   10 Z(I,J)=SUM
      RETURN
      END
C
      SUBROUTINE MULTRL(X,Y,Z,MM,NN,MJ1,MJ2)
C     TRANSPOSE MULTIPLY (LEFT)
C     Z=X'*Y
C     Z: SYMMETRIC
C     (LOWER TRIANGLE OF UPPER LEFT NN X NN OF Z)=(UPPER LEFT MM X NN OF
C     X)'*(UPPER LEFT MM X NN OF Y)
C     (MJ1,MJ2): ABSOLUTE DIMENSION OF X AND Y IN THE MAIN ROUTINE
C     (MJ2,MJ2): ABSOLUTE DIMENSION OF Z IN THE MAIN ROUTINE
      IMPLICIT REAL*8(A-H,O-Z)
      DIMENSION X(MJ1,MJ2),Y(MJ1,MJ2),Z(MJ2,MJ2)
      DO 10 I=1,NN
      DO 10 J=1,I
      SUM=0.0
      DO 11 K=1,MM
   11 SUM=SUM+X(K,I)*Y(K,J)
   10 Z(I,J)=SUM
      RETURN
      END
C
      SUBROUTINE REMTSB(X,MM,ISW,MJ)
C     SYMMETRIC MATRIX X(MM,MM) (ON AND BELOW DIAGONAL ELEMENTS) INPUT
C     ISW=1: ROWWISE INPUT
C     ISW=2: COLUMNWISE INPUT
C     UPPER TRIANGLE OF X(MM,MM) ARRANGEMENT
      IMPLICIT REAL*8(A-H,O-Z)
      DIMENSION X(MJ,MJ)
      GO TO(8,9),ISW
    8 DO 10 I=1,MM
   10 READ(5,2) (X(I,J),J=1,I)
      GO TO 12
    9 DO 11 J=1,MM
   11 READ(5,2) (X(I,J),I=J,MM)
   12 IF(MM.EQ.1) GO TO 14
      MM1=MM-1
      DO 13 I=1,MM1
      I1=I+1
      DO 13 J=I1,MM
   13 X(I,J)=X(J,I)
   14 RETURN
    2 FORMAT(8F10.4)
C     THE FOREGOING FORMAT SPECIFICATION SHOULD BE CHANGED TO MEET THE
C     INDIVIDUAL REQUIREMENTS OF SPECIFIC JOBS.
      END
```

```
      PROGRAM OPTSIM
C     PROGRAM 5.5.2   OPTIMAL CONTROL SIMULATION
C     THIS PROGRAM PERFORMS OPTIMAL CONTROL SIMULATION FOR THE
C     CONTROLLER DESIGNED BY PROGRAM 5.5.1 AND EVALUATES THE MEANS AND
C     VARIANCES OF THE CONTROLLED AND MANIPULATED VARIABLES X AND Y.
C     FOLLOWING CONSTANTS SHOULD BE PROVIDED BESIDES THE OUTPUTS OF
C     PROGRAM 5.5.1   OPTIMAL CONTROLLER DESIGN TO START THIS PROGRAM.
C     NS: NUMBER OF STEPS OF SIMULATION
C     INTP=1: TO SUPPRESS HISTORY OUTPUT
C     INTP=2: TO PRINT OUT THE HISTORY
C     THE SEQUENCE OF NS IR-DIMENSIONAL VECTORS W, REPRESENTING WHITE
C     NOISE (OR IMPULSE) IS ALSO REQUIRED AS INPUT.
      IMPLICIT REAL*8(A-H,O-Z)
      DIMENSION A(75,5),B(75,5),Q1(5,5),R(5,5),G(5,75)
      DIMENSION X(5),XS(5),XS2(5)
      DIMENSION Y(5),YS(5),YS2(5)
      DIMENSION W(5),Z(75),C(75)
      DIMENSION XMEAN(5),XS2MEA(5),XVAR(5)
      DIMENSION YMEAN(5),YS2MEA(5),YVAR(5)
C     ABSOLUTE DIMENSIONS USED FOR SUBROUTINE CALL
      MJ1=5
      MJ2=5
      MJ3=75
C     INITIAL CONDITION INPUT AND OUTPUT
      READ(5,1) NS,INTP
C     READING THE OUTPUTS OF PROGRAM 5.5.1 OPTDES
      READ(5,1) N,M,IR,L
      MR=M*IR
      CALL REMATX(Q1,IR,IR,1,MJ1,MJ1)
      CALL REMATX(R,L,L,1,MJ2,MJ2)
      CALL REMATX(A,MR,IR,1,MJ3,MJ1)
      CALL REMATX(B,MR,L,1,MJ3,MJ2)
      CALL REMATX(G,L,MR,1,MJ2,MJ3)
      WRITE(6,60)
      WRITE(6,61)
      WRITE(6,62) N,M,IR,L,NS
      WRITE(6,65)
      CALL SUBMPR(Q1,IR,IR,MJ1,MJ1)
      WRITE(6,66)
      CALL SUBMPR(R,L,L,MJ2,MJ2)
      WRITE(6,63)
      CALL SUBMPR(A,MR,IR,MJ3,MJ1)
      WRITE(6,64)
      CALL SUBMPR(B,MR,L,MJ3,MJ2)
      WRITE(6,67)
      CALL SUBMPR(G,L,MR,MJ2,MJ3)
C     INITIAL CONDITIONING
      DO 6 I=1,IR
      X(I)=0.0
      XS(I)=0.0
    6 XS2(I)=0.0
      DO 7 I=1,L
      Y(I)=0.0
      YS(I)=0.0
    7 YS2(I)=0.0
      DO 8 I=1,MR
    8 C(I)=0.0
      MR1=MR-IR
C     START OF SIMULATION
C     NOISE INPUT
      DO 10 INS=1,NS
      READ(5,3) (W(I),I=1,IR)
C     X COMPUTATION
      CALL VECADL(C,W,IR)
      DO 9 I=1,IR
    9 X(I)=C(I)
C     Y COMPUTATION
      CALL MULVER(G,C,Y,L,MR,MJ2,MJ3)
      IF(INS.EQ.NS) GO TO 101
      CALL MULVER(A,X,Z,MR,IR,MJ3,MJ1)
      IF(M.EQ.1) GO TO 360
      DO 20 I=1,MR1
      IPR=I+IR
   20 Z(I)=Z(I)+C(IPR)
  360 CALL MULVER(B,Y,C,MR,L,MJ3,MJ2)
      CALL VECADL(C,Z,MR)
```

```
C     SUM AND SUM OF SQUARES COMPUTATION
201 CALL VECADL(XS,X,IR)
    CALL VECADL(YS,Y,L)
    DO 30 I=1,IR
 30 XS2(I)=XS2(I)+X(I)**2
    DO 31 I=1,L
 31 YS2(I)=YS2(I)+Y(I)**2
    IF(INTP.EQ.1) GO TO 10
C     X,Y,W PRINT OUT
    WRITE(6,260) INS
    WRITE(6,261)
    CALL PRCOL2(X,W,1,IR,0)
    WRITE(6,262)
    CALL PRCOL1(Y,1,L,0)
 10 CONTINUE
C     MEAN, MEAN SQUARE AND VARIANCE COMPUTATION
    ANS=NS
    BNS=1.0/ANS
    DO 40 I=1,IR
    XMEAN(I)=BNS*XS(I)
    XS2MEA(I)=BNS*XS2(I)
 40 XVAR(I)=XS2MEA(I)-XMEAN(I)**2
    DO 41 I=1,L
    YMEAN(I)=BNS*YS(I)
    YS2MEA(I)=BNS*YS2(I)
 41 YVAR(I)=YS2MEA(I)-YMEAN(I)**2
    WRITE(6,160)
    WRITE(6,161)
    CALL PRCOL4(XMEAN,XS2,XS2MEA,XVAR,1,IR,0)
    WRITE(6,163)
    CALL PRCOL4(YMEAN,YS2,YS2MEA,YVAR,1,L,0)
    STOP
  1 FORMAT(10I5)
  3 FORMAT(6D12.3)
 60 FORMAT(1H1,42HPROGRAM 5.5.2   OPTIMAL CONTROL SIMULATION)
 61 FORMAT(1H0,17HINITIAL CONDITION.
 62 FORMAT(1H0,2HN=,I5,5X,2HM=,I5,5X,3HIR=,I5,5X,2HL=,I5,5X,3HNS=,I5)
 63 FORMAT(//1H0,44HFIRST IR COLUMNS OF TRANSITION MATRIX (AI'S))
 64 FORMAT(//1H0,19HGAMMA MATRIX (BI'S))
 65 FORMAT(//1H0,7HA1(I,J))
 66 FORMAT(//1H0,6HB(I,J))
 67 FORMAT(//1H0,13HGAIN MATRIX G)
160 FORMAT(//1H0,29X,4HX(I))
161 FORMAT(1H0,4X,1HI,5X,9HMEAN OF X,5X,11HSUM OF X**2,3X,12HMEAN OF X
   A**2,2X,13HVARIANCE OF X)
163 FORMAT(//1H0,29X,4HY(I))
164 FORMAT(1H0,4X,1HI,5X,9HMEAN OF Y,5X,11HSUM OF Y**2,3X,12HMEAN OF Y
   A**2,2X,13HVARIANCE OF Y)
260 FORMAT(/1H0,4HINS=,I5)
261 FORMAT(1H0,4X,1HI,5X,4HX(I),10X,4HW(I))
262 FORMAT(1H0,4X,1HI,5X,4HY(I))
    END
C
    SUBROUTINE MULVER(X,Y,Z,MM,NN,MJ1,MJ2)
C     Z=X*Y (X: MATRIX  Y,Z: VECTORS)
C     (MJ1,MJ2): ABSOLUTE DIMENSION OF X IN THE MAIN ROUTINE
    IMPLICIT REAL*8(A-H,O-Z)
    DIMENSION X(MJ1,MJ2),Y(NN),Z(MM)
    DO 10 I=1,MM
    SUM=0.0
    DO 11 J=1,NN
 11 SUM=SUM+X(I,J)*Y(J)
 10 Z(I)=SUM
    RETURN
    END
C
    SUBROUTINE VECADL(X,Y,MM)
C     X=X+Y (X,Y: VECTORS)
    IMPLICIT REAL*8(A-H,O-Z)
    DIMENSION X(MM),Y(MM)
    DO 10 I=1,MM
 10 X(I)=X(I)+Y(I)
    RETURN
    END
```

```
      PROGRAM WNOISE
C     PROGRAM 5.5.3   WHITE NOISE GENERATOR
C     THIS PROGRAM GENERATES APPROXIMATELY GAUSSIAN VECTOR WHITE NOISE
C     TO BE USED AS INPUT N OF PROGRAM 5.5.2 OPTSIM.
C     ON TOP OF THE OUTPUT CF PROGRAM 5.3.2 FPEC, WHICH IS TO BE USED
C     AS INPUT TO PROGRAM 5.5.2 OPTSIM, ONE CARD WITH SPECIFICATION
C     OF THE LENGTH NRA OF WHITE NOISE RECORD TO BE GENERATED SHOULD
C     BE ADDED TO FORM THE INPUT TO THIS PROGRAM.
C     NRA: LENGTH OF WHITE NOISE RECORD TO BE GENERATED
      IMPLICIT REAL*8(A-H,O-Z)
      REAL*4 RANDOM
      DIMENSION SD(5,5),A(10),Y(5),Z(5)
      DIMENSION X1(100,5)
C     ABSOLUTE DIMENSIONS USED FOR SUBROUTINE CALL
      MJO=100
      MJ=5
C     NRA SPECIFICATION
      READ(5,1) NRA
C     READING THE OUTPUTS OF PROGRAM 5.3.2 FPEC
      READ(5,1) N,M,IR,IL
      CALL REMATX(SD,IR,IR,1,MJ,MJ)
C     FOLLOWING INPUT IS NONEFFECTIVE.
      IP=IR+IL
      MR=M*IR
      DO 8 JJ=1,MR
    8 READ(5,2) (A(II),II=1,IP)
      WRITE(6,60)
      WRITE(6,61)
      WRITE(6,62) NRA,N,M,IR
      WRITE(6,63)
      CALL SUBMPR(SD,IR,IR,MJ,MJ)
      WRITE(6,100)
C     MATRIX L COMPUTATION
      CALL LTINV(SD,IR,MJ)
C     MATRIX L ARRANGEMENT
      IF(IR.EQ.1) GO TO 260
      DO 12 I=2,IR
      IM1=I-1
      DO 12 J=1,IM1
   12 SD(I,J)=SD(J,I)
C     RANDOM NUMBER GENERATION
  260 RC=4.0/3.0
      RCONST=DSQRT(RC)
      XX=RANDOM(1)
      IND=99
      IND1=IND+1
      I1=0
      I2=0
      IIC=0
  160 I1=I2+1
      I2=I1+IND
      IF(I2.LE.NRA) GO TO 130
      I2=NRA
  130 DO 14 I=I1,I2
      II=I-IIC
      DO 15 J=1,IR
      SUM=0.0
      DO 20 JJ=1,9
   20 SUM=SUM+RANDOM(0)
      SUM=SUM-4.5
   15 X1(II,J)=SUM*RCONST
   14 CONTINUE
```

```
C       WHITE NOISE GENERATION
        DO 16 I=I1,I2
        II=I-IIC
        DO 17 J=1,IR
     17 Y(J)=X1(II,J)
        CALL LTRVFC(SD,Y,Z,IR,IR,MJ,MJ)
        DO 18 J=1,IR
     18 X1(II,J)=Z(J)
     16 CONTINUE
C       WHITE NOISE PRINT AND PUNCH OUT
        WRITE(6,101) IIC
        I3=I2-IIC
        CALL SUBMPR(X1,I3,IR,MJC,MJ)
        DO 40 I=I1,I2
        II=I-IIC
     40 WRITE(7,3) (X1(II,J),J=1,IR)
        IIC=IIC+IND1
        IF(I2.LT.NRA) GO TO 160
        STOP
      1 FORMAT(10I5)
      2 FORMAT(4D20.10)
      3 FORMAT(6D12.3)
     60 FORMAT(1H1,27HPROGRAM 5.5.3    WHITE NOISE)
     61 FORMAT(1H0,17HINITIAL CONDITION)
     62 FORMAT(1H0,4HNRA=,I5,5X,2HN=,I5,5X,2HM=,I5,5X,3HIR=,I5)
     63 FORMAT(/1H0,7HSD(I,J))
    100 FORMAT(////1H0,11HWHITE NOISE)
    101 FORMAT(1H0,4HIIC=,I5,5X,11HX1(IIC+I,J))
        END
C
        SUBROUTINE LTINV(R,K,MJ)
C       THIS SUBROUTINE FACTORIZES (R(I,J): I,J=1,K) INTO R=L*L',
C       WITH L LOWER TRIANGLE, AND GIVE L' ON AND ABOVE DIAGONAL OF R.
C       MJ: ABSOLUTE DIMENSION OF R IN THE MAIN ROUTINE
        IMPLICIT REAL*8(A-H,O-Z)
        DIMENSION R(MJ,MJ)
        DO 10 L=1,K
        RPIVOT=1.0/DSQRT(R(L,L))
        R(L,L)=1.0/RPIVOT
        DO 12 I=1,K
        IF(I.EQ.L) GO TO 12
        R(L,I)=RPIVOT*R(L,I)
     12 CONTINUE
        IF(L.EQ.K) GO TO 11
        L1=L+1
        DO 13 I=L1,K
        RIL=-RPIVOT*R(I,L)
        R(I,L)=RIL*RPIVOT
        DO 14 M=1,K
        IF(M.EQ.L) GO TO 14
        R(I,M)=R(I,M)+RIL*R(L,M)
     14 CONTINUE
     13 CONTINUE
     10 CONTINUE
     11 RETURN
        END
```

```
      SUBROUTINE LTRVEC(X,Y,Z,MM,NN,MJ1,MJ2)
C     Z=X*Y
C     (VECTOR Z)=(LOWER TRIANGLE OF UPPER LEFT MM X NN OF X)*(VECTOR Y)
C     (MJ1,MJ2): ABSOLUTE DIMENSION OF X IN THE MAIN ROUTINE.
      IMPLICIT REAL*8(A-H,O-Z)
      DIMENSION X(MJ1,MJ2),Y(NN),Z(MM)
      DO 10 I=1,MM
      SUM=0.0
      DO 11 J=1,I
   11 SUM=SUM+X(I,J)*Y(J)
   10 Z(I)=SUM
      RETURN
      END
C
      FUNCTION RANDOM(K)
C     RANDOM NUMBER GENERATOR
      MCST11=11
      MC100=100
      IF(K) 1,2,1
C     STARTING NUMBER FOR GENERATOR
    1 K1=53
      K2=95
      K3=27
      K4=04
      WRITE(6,4) K1,K2,K3,K4
    2 M1=MCST11*K4
      M2=MCST11*K3
      M3=MCST11*K2+K4
      M4=MCST11*K1+K3
      J=M1/MC100
      K4=M1-MC100*J
      M2=M2+J
      J=M2/MC100
      K3=M2-MC100*J
      M3=M3+J
      J=M3/MC100
      K2=M3-MC100*J
      M4=M4+J
      J=M4/MC100
      K1=M4-MC100*J
      X1=K1
      X2=K2
      RANDOM=X1*1.E-2+X2*1.E-4
      RETURN
    4 FORMAT(1H0,29HSTARTING NUMBER FOR GENERATOR,5X,4I3)
      END
```

5.6 Subroutines

Here, we collectively present subroutines that have been used in more than one program. The subroutines are placed in alphabetical order. The function of each subroutine is explained by the comment given in the main program or by the comment within the subroutine itself. For the subroutines INVDET and REARRA, the variables must be properly defined when these subroutines are used for complex computations.

```
      SUBROUTINE AUSP(FC,P1,LAGH1,A,LA1)
C     THIS SUBROUTINE COMPUTES SMOOTHED AUTO SPECTRUM.
C     FC: OUTPUT OF FGERCO
C     P1: SMOOTHED SPECTRUM
C     LAGH1: DIMENSION OF FC AND P1
C     A: SMOOTHING COEFFICIENTS
C     LA1: DIMENSION OF A (LESS THAN 11)
      IMPLICIT REAL*8(A-H,C-Z)
      DIMENSION FC(LAGH1),P1(LAGH1),A(LA1)
      DIMENSION FC1(521)
      LA=LA1-1
      LAGSHF=LAGH1+2*LA
C     FC SHIFT-RIGHT BY LA FOR END CORRECTION
      CALL ECORCO(FC,LAGH1,FC1,LAGSHF,LA1)
C     SMOOTHING
      CALL SMOSPE(FC1,LAGSHF,A,LA1,P1,LAGH1)
      RETURN
      END


      SUBROUTINE CORNOM(C,CN,LAGH1,CX0,CY0)
C     NORMALIZATION OF COVARIANCE
      IMPLICIT REAL*8(A-H,O-Z)
      DIMENSION C(LAGH1),CN(LAGH1)
      DS=1.0/DSQRT(CX0*CY0)
      DO 10 I=1,LAGH1
   10 CN(I)=C(I)*DS
      RETURN
      END


      SUBROUTINE CROSCO(X,Y,N,C,LAGH1)
C     THIS SUBROUTINE COMPUTES C(L)=COVARIANCE(X(S+L),Y(S))
C     (L=0,1,...,LAGH1-1).
      DOUBLE PRECISION C,T
      DIMENSION X(N),Y(N),C(LAGH1)
      AN=N
      BN=1.0/AN
      DO 10 II=1,LAGH1
      I=II-1
      T=0.0
      IL=N-I
      DO 20 J=1,IL
      J1=J+I
   20 T=T+DBLE(X(J1))*Y(J)
   10 C(II)=T*BN
      RETURN
      END


      SUBROUTINE ECORCO(FC,LAGH1,FC1,LAGSHF,LA1)
C     FC SHIFT-RIGHT BY LA FOR REAL PART END CORRECTION
      IMPLICIT REAL*8(A-H,O-Z)
      DIMENSION FC(LAGH1),FC1(LAGSHF)
      LAGH2=LAGH1+1
      LA=LA1-1
      DO 100 I=1,LAGH1
      I1=LAGH2-I
      I2=I1+LA
  100 FC1(I2)=FC(I1)
      LA2=LAGH1+LA
      DO 110 I=1,LA
      I1=LA1-I
      I2=LA1+I
      I3=LA2-I
      I4=LA2+I
      FC1(I1)=FC1(I2)
  110 FC1(I4)=FC1(I3)
      RETURN
      END
```

```
      SUBROUTINE FGERCO(G,LGP1,FC,LF1)
C     FOURIER TRANSFORM (GOERTZEL METHOD)
C     THIS SUBROUTINE COMPUTES FOURIER TRANSFORM OF G(I),I=0,1,...,LG AT
C     FREQUENCIES K/(2*LF),K=0,1,...,LF AND RETURNS COSIN TRANSFORM IN
C     FC(K).
      IMPLICIT REAL*8(A-H,O-Z)
      DIMENSION G(LGP1),FC(LF1)
      LG=LGP1-1
      LF=LF1-1
C     REVERSAL OF G(I),I=1,...,LGP1 INTO G(LG3-I)    LG3=LGP1+1
      IF(LGP1.LE.1) GO TO 110
      LG3=LGP1+1
      LG4=LGP1/2
      DO 100 I=1,LG4
      I2=LG3-I
      T=G(I)
      G(I)=G(I2)
  100 G(I2)=T
  110 PI=3.1415926536
      ALF=LF
      T=PI/ALF
      DO 10 K=1,LF1
      AK=K-1
      TK=T*AK
      CK=DCOS(TK)
      CK2=CK+CK
      UM2=0.0
      UM1=0.0
      IF(LG.EQ.0) GO TO 12
      DO 11 I=1,LG
      UMO=CK2*UM1-UM2+G(I)
      UM2=UM1
   11 UM1=UMO
   12 FC(K)=CK*UM1-UM2+G(LGP1)
   10 CONTINUE
      RETURN
      END


      SUBROUTINE FGER1
C     FOURIER TRANSFORM(GOERTZEL METHOD)
C     THIS SUBROUTINE COMPUTES ONE VALUE OF THE FOURIER TRANSFORM BY
C     GOERTZEL METHOD.
      IMPLICIT REAL*8(A-H,O-W)
      INTEGER H
      COMMON G,LG,GR,GI,H,JJF
      DIMENSION G(31)
      LGP1=LG+1
C     REVERSAL OF G(I),I=1,...,LGP1 INTO G(LG3-I)    LG3=LGP1+1
      IF(LGP1.LE.1) GO TO 110
      LG3=LGP1+1
      LG4=LG3/2
      DO 100 I=1,LG4
      I2=LG3-I
      T=G(I)
      G(I)=G(I2)
  100 G(I2)=T
  110 PI=3.1415926536
      AH=H
      T=PI/AH
      AK=JJF-1
      TK=T*AK
      CK=DCOS(TK)
      SK=DSIN(TK)
      CK2=CK+CK
      UM2=0.0
      UM1=0.0
      IF(LG.EQ.0) GO TO 12
      DO 11 I=1,LG
      UMO=CK2*UM1-UM2+G(I)
      UM2=UM1
   11 UM1=UMO
   12 GR=CK*UM1-UM2+G(LGP1)
      GI=-SK*UM1
      RETURN
      END
```

```
      SUBROUTINE INVDET(X,XDET,MM,MJ)
C     THIS SUBROUTINE COMPUTES THE INVERSE AND DETERMINANT OF
C     UPPER LEFT MM X MM OF X.
C     X: ORIGINAL MATRIX
C     MM: DIMENSION OF UPPER LEFT OF X (SHOULD BE LESS THAN 11)
C     XDET: DETERMINANT OF UPPER LEFT MM X MM OF X
C     MJ: ABSOLUTE DIMENSION OF X IN THE MAIN ROUTINE
C     THE INVERSE MATRIX IS OVERWRITTEN ON THE ORIGINAL.
C     NEXT STATEMENT SHOULD BE REPLACED BY
C     IMPLICIT COMPLEX*16(X)
C     FOR COMPLEX VERSION.  ALSO STATEMENT NO.1 NEEDS MODIFICATION.
      IMPLICIT REAL*8(X)
      DIMENSION X(MJ,MJ)
      DIMENSION IDS(10)
      XDET=1.0
      DO 10 L=1,MM
C     PIVOTING AT L-TH STAGE
      XMAXP=0.10000D-10
      MAXI=0
      DO 110 I=L,MM
C     FOR COMPLEX VERSION NEXT STATEMENT SHOULD BE REPLACED BY
C     IF(CDABS(XMAXP).GE.CDABS(X(I,L))) GO TO 110
    1 IF(DABS(XMAXP).GE.DABS(X(I,L))) GO TO 110
      XMAXP=X(I,L)
      MAXI=I
  110 CONTINUE
      IDS(L)=MAXI
      IF(MAXI.EQ.L) GO TO 120
      IF(MAXI.GT.0) GO TO 121
      XDET=0.0
      GO TO 140
C     ROW INTERCHANGE
  121 DO 14 J=1,MM
      XC=X(MAXI,J)
      X(MAXI,J)=X(L,J)
   14 X(L,J)=XC
      XDET=-XDET
  120 XDET=XDET*XMAXP
      XC=1.0/XMAXP
      X(L,L)=1.0
      DO 11 J=1,MM
   11 X(L,J)=X(L,J)*XC
      DO 12 I=1,MM
      IF(I.EQ.L) GO TO 12
      XC=X(I,L)
      X(I,L)=0.0
      DO 13 J=1,MM
   13 X(I,J)=X(I,J)-XC*X(L,J)
   12 CONTINUE
   10 CONTINUE
      IF(MN.GT.1) GO TO 123
      GO TO 140
C     COLUMN INTERCHANGE
  123 MM1=MM-1
      DO 130 J=1,MM1
      MMJ=MM-J
      JJ=IDS(MMJ)
      IF(JJ.EQ.MMJ) GO TO 130
      DO 131 I=1,MM
      XC=X(I,JJ)
      X(I,JJ)=X(I,MMJ)
  131 X(I,MMJ)=XC
  130 CONTINUE
  140 RETURN
      END


      SUBROUTINE MATADL(X,Y,MM,NN,MJ1,MJ2)
C     MATRIX ADDITION.
C     X=X+Y
C     (UPPER LEFT MM X NN OF X)=(UPPER LEFT MM X NN OF X)+(UPPER LEFT
C     MM X NN OF Y).
C     (MJ1,MJ2): ABSOLUTE DIMENSION OF X AND Y IN THE MAIN ROUTINE
      IMPLICIT REAL*8(A-H,O-Z)
      DIMENSION X(MJ1,MJ2),Y(MJ1,MJ2)
      DO 10 I=1,MM
      DO 10 J=1,NN
   10 X(I,J)=X(I,J)+Y(I,J)
      RETURN
      END
```

```
      SUBROUTINE  MULPLY(X,Y,Z,MM,NC,NC,MJ1,MJ2,MJ3)
C     MATRIX MULTIPLICATION.
C     Z=X*Y
C     (UPPER LEFT MM X NC OF Z)=(UPPER LEFT M X NN OF X)*(UPPER LEFT
C     NN X NC OF Y).
C     (MJ1,MJ2): ABSOLUTE DIMENSION OF X IN THE MAIN ROUTINE
C     (MJ2,MJ3): ABSOLUTE DIMENSION OF Y IN THE MAIN ROUTINE
C     (MJ1,MJ3): ABSOLUTE DIMENSION OF Z IN THE MAIN ROUTINE
      IMPLICIT REAL*8(A-H,O-Z)
      DIMENSION X(MJ1,MJ2),Y(MJ2,MJ3),Z(MJ1,MJ3)
      DO 10 I=1,MM
      DO 11 J=1,NC
      SUM=0.0
      DO 12 K=1,NN
   12 SUM=SUM+X(I,K)*Y(K,J)
      Z(I,J)=SUM
   11 CONTINUE
   10 CONTINUE
      RETURN
      END


      SUBROUTINE PRCOL1(P1,INDI,INDL,ISHIFT)
C     THIS SUBROUTINE PRINTS VECTOR (P1(I),I=INDI,INDL)
C     COLUMNWISE IN THE FORMAT (I-ISHIFT,P1(I),I=INDI,INDL).
      IMPLICIT REAL*8(A-H,O-Z)
      DIMENSION P1(INDL)
      DO 61 I=INDI,INDL
      IM1=I-ISHIFT
   61 WRITE(6,62) IM1,P1(I)
      RETURN
   62 FORMAT(1H ,I5,D16.5)
      END


      SUBROUTINE PRCOL2(P1,P2,INDI,INDL,ISHIFT)
C     THIS SUBROUTINE PRINTS VECTORS
C     (P1(I),P2(I),I=INDI,INDL)
C     COLUMNWISE IN THE FORMAT
C     (I-ISHIFT,P1(I),P2(I),I=INDI,INDL).
      IMPLICIT REAL*8(A-H,O-Z)
      DIMENSION P1(INDL),P2(INDL)
      DO 61 I=INDI,INDL
      IM1=I-ISHIFT
   61 WRITE(6,62) IM1,P1(I),P2(I)
      RETURN
   62 FORMAT(1H ,I5,2X,2D14.5)
      END


      SUBROUTINE PRCOL3(P1,P2,P3,INDI,INDL,ISHIFT)
C     THIS SUBROUTINE PRINTS OUT VECTORS
C     (P1(I),P2(I),P3(I),I=INDI,INDL).
C     COLUMNWISE IN THE FORMAT
C     (I-ISHIFT,P1(I),P2(I),P3(I),I=INDI,INDL)
      IMPLICIT REAL*8(A-H,O-Z)
      DIMENSION P1(INDL),P2(INDL),P3(INDL)
      DO 61 I=INDI,INDL
      IM1=I-ISHIFT
   61 WRITE(6,62) IM1,P1(I),P2(I),P3(I)
      RETURN
   62 FORMAT(1H ,I5,2X,3D14.5)
      END


      SUBROUTINE PRCOL4(P1,P2,P3,P4,INDI,INDL,ISHIFT)
C     THIS SUBROUTINE PRINTS VECTORS
C     (P1(I),P2(I),P3(I),P4(I),I=INDI,INDL)
C     COLUMNWISE IN THE FORMAT
C     (I-ISHIFT,P1(I),P2(I),P3(I),P4(I),I=INDI,INDL).
      IMPLICIT REAL*8(A-H,O-Z)
      DIMENSION P1(INDL),P2(INDL),P3(INDL),P4(INDL)
      DO 61 I=INDI,INDL
      IM1=I-ISHIFT
   61 WRITE(6,62) IM1,P1(I),P2(I),P3(I),P4(I)
      RETURN
   62 FORMAT(1H ,I5,2X,4D14.5)
      END
```

```
      SUBROUTINE PRCCL6(P1,P2,P3,P4,P5,P6,INDI,INDL,ISHIFT)
C     THIS SUBROUTINE PRINTS VECTORS
C     (P1(I),P2(I),P3(I),P4(I),P5(I),P6(I),I=INDI,INDL)
C     COLUMNWISE IN THE FORMAT
C     (I-ISHIFT,P1(I),P2(I),P3(I),P4(I),P5(I),P6(I),I=INDI,INDL).
      IMPLICIT REAL*8(A-H,O-Z)
      DIMENSION P1(INDL),P2(INDL),P3(INDL),P4(INDL),P5(INDL),P6(INDL)
      DO 61 I=INDI,INDL
      IM1=I-ISHIFT
   61 WRITE(6,62) IM1,P1(I),P2(I),P3(I),P4(I),P5(I),P6(I)
      RETURN
   62 FORMAT(1H ,I5,2X,6D14.5)
      END


      SUBROUTINE PRCPMA(X,MC,LC,MJ1,MJ2)
C     THIS SUBROUTINE PRINTS OUT UPPER LEFT MC X LC OF COMPLEX MATRIX X.
C     (MJ1,MJ2): ABSOLUTE DIMENSION OF X IN THE MAIN ROUTINE
      IMPLICIT COMPLEX*16(X)
      DIMENSION X(MJ1,MJ2)
      WRITE(6,39) MC,LC
      WRITE(6,20)
      WRITE(6,120)
      DO 10 I=1,MC
      J1=0
      J2=0
      JC=0
   16 J1=J2+1
      J2=J1+4
      IF(J2.LE.LC) GO TO 13
      J2=LC
   13 JC=JC+1
      IF(JC.GT.1) GO TO 19
      WRITE(6,21) I,(X(I,J),J=J1,J2)
      GO TO 17
   19 WRITE(6,22) (X(I,J),J=J1,J2)
   17 IF(J2.LT.LC) GO TO 16
   10 CONTINUE
      RETURN
   39 FORMAT(1H0,14HCOMPLEX MATRIX,5X,I5,4X,1HX,I5)
   20 FORMAT(1H0,18X,1H1,23X,1H2,23X,1H3,23X,1H4,23X,1H5)
  120 FORMAT(1H ,5X,5(4X,9HREAL PART,2X,10HIMAG. PART))
   21 FORMAT(1H0,I5,5(1X,2D12.5))
   22 FORMAT(1H ,5X,5(1X,2D12.5))
      END


      SUBROUTINE PRMAT3(X1,MM,NN,NC,ISHIFT,MJ0,MJ1,MJ2)
C     3-DIMENSIONAL MATRIX X1(MM,NN,NC) PRINT OUT
C     (MJ0,MJ1,MJ2): ABSOLUTE DIMENSION OF X1 IN THE MAIN ROUTINE
C     NN,NC: SHOULD BE LESS THAN 11
      IMPLICIT REAL*8(A-H,O-Z)
      DIMENSION X1(MJ0,MJ1,MJ2)
      DIMENSION X(10,10)
      WRITE(6,60) MM
      DO 61 I=1,MM
      DO 62 II=1,NN
      DO 62 JJ=1,NC
   62 X(II,JJ)=X1(I,II,JJ)
      I1=I-ISHIFT
      WRITE(6,63) I1
      CALL SUBMPR(X,NN,NC,MJ1,MJ2)
   61 CONTINUE
      RETURN
   60 FORMAT(1H0,3HLL=,I5)
   63 FORMAT(/1H0,2HI=,I5)
      END
```

```
      SUBROUTINE REARRA(X,INW,IPO,IP,MJ)
C     SUBMATRIX REARRANGEMENT
C     X: ORIGINAL MATRIX
C     INW: INDICATOR OF ADOPTED ROWS
C     IPO: DIMENSION OF ORIGINAL MATRIX, SHOULD BE LESS THAN 11
C     IP: DIMENSION OF REARRANGED SUBMATRIX
C     MJ: ABSOLUTE DIMENSION OF X IN THE MAIN ROUTINE
C     THE REARRANGED SUBMATRIX IS OVERWRITTEN ON THE ORIGINAL.
C     NEXT STATEMENT SHOULD BE REPLACED BY
C     IMPLICIT COMPLEX*16(X)
C     FOR COMPLEX VERSION.
      IMPLICIT REAL*8(X)
      DIMENSION X(MJ,MJ),INW(IP)
      DIMENSION IOD(10)
      DO 300 I=1,IPO
  300 IOD(I)=I
      DO 301 I=1,IP
      I1=INW(I)
      I2=IOD(I1)
      IF(I.EQ.I2) GO TO 301
C     ROW INTERCHANGE
      DO 312 JJ=1,IPO
      XC=X(I,JJ)
      X(I,JJ)=X(I2,JJ)
  312 X(I2,JJ)=XC
C     COLUMN INTERCHANGE
      DO 314 II=1,IPO
      XC=X(II,I)
      X(II,I)=X(II,I2)
  314 X(II,I2)=XC
      ID=IOD(I)
      IOD(I2)=ID
      IOD(ID)=I2
  301 CONTINUE
      RETURN
      END


      SUBROUTINE REMATX(X,MM,NN,ISW,MJ1,MJ2)
C     MATRIX X(MM,NN) INPUT
C     (MJ1,MJ2): ABSOLUTE DIMENSION OF X IN THE MAIN ROUTINE
C     ISW=1: ROWWISE INPUT
C     ISW=2: COLUMNWISE INPUT
      IMPLICIT REAL*8(A-H,O-Z)
      DIMENSION X(MJ1,MJ2)
      GO TO(8,9),ISW
    8 DO 10 I=1,MM
   10 READ(5,2) (X(I,J),J=1,NN)
      GO TO 12
    9 DO 11 J=1,NN
   11 READ(5,2) (X(I,J),I=1,MM)
   12 RETURN
    2 FORMAT(4D20.10)
C     THE FOREGOING FORMAT SPECIFICATION SHOULD BE CHANGED TO MEET THE
C     INDIVIDUAL REQUIREMENTS OF SPECIFIC JOBS.
      END


      SUBROUTINE REMAT3(X1,MM,NN,NC,ISW,MJO,MJ1,MJ2)
C     3-DIMENSIONAL MATRIX X1(MM,NN,NC) INPUT
C     ISW=1: ROWWISE INPUT
C     ISW=2: COLUMNWISE INPUT
C     (MJO,MJ1,MJ2): ABSOLUTE DIMENSION OF X1 IN THE MAIN ROUTINE
      IMPLICIT REAL*8(A-H,O-Z)
      DIMENSION X1(MJO,MJ1,MJ2)
      GO TO(8,9),ISW
    8 DO 10 I=1,MM
      DO 110 II=1,NN
  110 READ(5,2) (X1(I,II,JJ),JJ=1,NC)
   10 CONTINUE
      GO TO 12
    9 DO 11 I=1,MM
      DO 111 JJ=1,NC
  111 READ(5,2) (X1(I,II,JJ),II=1,NN)
   11 CONTINUE
   12 RETURN
    2 FORMAT(4D20.10)
C     THE FOREGOING FORMAT SPECIFICATION SHOULD BE CHANGED TO MEET THE
C     INDIVIDUAL REQUIREMENTS OF SPECIFIC JOBS.
      END
```

```
      SUBROUTINE SIGNIF(P1,P2,P3,LAGH1,N)
C     SIGNIFICANCE TEST
C     P1: SPECTRUM SMOOTHED BY WINDOW W1
C     P2: SPECTRUM SMOOTHED BY WINDOW W2
C     P3: TEST STATISTICS
C     LAGH1: DIMENSION OF PI (I=1,2,3)
C     N: LENGTH OF THE ORIGINAL DATA
      IMPLICIT REAL*8(A-H,O-Z)
      DIMENSION P1(LAGH1),P2(LAGH1),P3(LAGH1)
      LAGH=LAGH1-1
      H=LAGH
      AN=N
      HAN=H/AN
      SD2=0.43*DSQRT(HAN)
      SD3=1.0/SD2
      DO 10 I=1,LAGH1
      T=P2(I)/P1(I)-1.0
   10 P3(I)=DABS(T)*SD3
      RETURN
      END


      SUBROUTINE SMEADL(X,N,XMEAN)
C     MEAN DELETION
      DIMENSION X(N)
      AN=N
      XMEAN=SUMF(X,N)/AN
      DO 10 I=1,N
   10 X(I)=X(I)-XMEAN
      RETURN
      END


      FUNCTION SUMF(X,N)
      DIMENSION X(N)
      SUMF=0.0
      DO 10 I=1,N
   10 SUMF=SUMF+X(I)
      RETURN
      END

      SUBROUTINE SMOSPE(X,LAGSHF,A,LA1,Z,LAGH1)
C     SPECTRUM SMOOTHING BY THE FORMULA
C     Z(I)=A(0)X(I)+A(1)(X(I+1)+X(I-1))+...+A(LA)(X(I+LA)+X(I-LA))
C     I=0,1,...,LAGH.
C     ACTUAL X(I) IS SHIFTED TO THE RIGHT BY LA FOR END CORRECTION.
      IMPLICIT REAL*8(A-H,O-Z)
      DIMENSION X(LAGSHF),A(LA1),Z(LAGH1)
      LA=LA1-1
      DO 10 I=1,LAGH1
      I0=I+LA
      SUM1=0.0
      DO 11 J=1,LA
      J1=I0-J
      J2=I0+J
   11 SUM1=SUM1+A(J+1)*(X(J1)+X(J2))
   10 Z(I)=A(1)*X(I0)+SUM1
      RETURN
      END
```

```
      SUBROUTINE SUBD12(N,LAGH,K,D1,D2)
C     CONSTANTS D1,D2 COMPUTATION
      IMPLICIT REAL*8(A-H,O-Z)
C     L1: NUMBER OF A(I)S (LESS THAN 5)
      DIMENSION A(4)
      L1=2
      A(1)=0.5
      A(2)=0.25
      AN=N
      H=LAGH
      SUM=0.0
      DO 20 I=2,L1
   20 SUM=SUM+A(I)**2
      SUM=SUM+SUM+A(1)**2
      SUM=SUM+SUM
      NF=AN/(H*SUM)+0.5
      FK=NF-K
      IF(FK.EQ.0.0) GO TO 100
      C1=FK-1.40
      IF(C1.EQ.0.0) GO TO 100
      D1=(3.84+10.0/C1)/FK
      IF(D1.LT.0.0) GO TO 100
      D1=DSQRT(D1)
      GO TO 110
  100 D1=100.0
  110 C2=FK+FK-1.40
      IF(C2.EQ.0.0) GO TO 120
      D2=(3.0+10.0/C2)/FK
      IF(D2.LT.0.0) GO TO 120
      D2=DSQRT(D2)
      GO TO 130
  120 D2=100.
  130 RETURN
      END


      SUBROUTINE SUBMPR(RM,MR,LC,MJ1,MJ2)
C     THIS SUBROUTINE PRINTS OUT UPPER LEFT MR X LC OF RM.
C     (MJ1,MJ2): ABSOLUTE DIMENSION OF X IN THE MAIN ROUTINE
      IMPLICIT REAL*8(A-H,O-Z)
      DIMENSION RM(MJ1,MJ2)
      WRITE(6,39) MR,LC
      WRITE(6,20)
      DO 10 I=1,MR
      J1=0
      J2=0
      JC=0
   16 J1=J2+1
      J2=J1+9
      IF(J2.LE.LC) GO TO 13
      J2=LC
   13 JC=JC+1
      IF(JC.GT.1) GO TO 19
   18 WRITE(6,21) I,(RM(I,J),J=J1,J2)
      GO TO 17
   19 WRITE(6,22) (RM(I,J),J=J1,J2)
   17 IF(J2.LT.LC) GO TO 16
   10 CONTINUE
      RETURN
   39 FORMAT(1H0,6HMATRIX,5X,I5,4X,1HX,I5)
   20 FORMAT(1H0,15X,1H1,11X,1H2,11X,1H3,11X,1H4,11X,1H5,11X,1H6,11X,1
     AH7,11X,1H8,11X,1H9,10X,2H10)
   21 FORMAT(1H0,I5,4X,10D12.5)
   22 FORMAT(1H ,9X,10D12.5)
      END
```

```
      SUBROUTINE TRAMOL(X,Y,Z,MM,NN,NC,MJ1,MJ2,MJ3)
C     TRANSPOSE MULTIPLY (LEFT)
C     Z=X'*Y
C     (UPPER LEFT NN X NC OF Z)=(UPPER LEFT MM X NN OF X)'*(UPPER LEFT
C     MM X NC OF Y).
C     (MJ1,MJ2): ABSOLUTE DIMENSION OF X IN THE MAIN ROUTINE
C     (MJ1,MJ3): ABSOLUTE DIMENSION OF Y IN THE MAIN ROUTINE
C     (MJ2,MJ3): ABSOLUTE DIMENSION OF Z IN THE MAIN ROUTINE
      IMPLICIT REAL*8(A-H,O-Z)
      DIMENSION X(MJ1,MJ2),Y(MJ1,MJ3),Z(MJ2,MJ3)
      DO 10 I=1,NN
      DO 11 J=1,NC
      SUM=0.0
      DO 12 K=1,MM
   12 SUM=SUM+X(K,I)*Y(K,J)
      Z(I,J)=SUM
   11 CONTINUE
   10 CONTINUE
      RETURN
      END
```

REFERENCES

[1] W. M. Gentleman and G. Sande, Fast Fourier Transform-for fun and profit, Fall Joint Computer Conference, AFIPS Proc., Spartan, Washington D.C. (1966) 563–578.
[2] R. W. Hamming, *Numerical Methods for Scientists and Engineers*, McGraw Hill, New York (1962).
[3] P.R. Lohnes and W. W. Cooley, *Introduction to Statistical Procedures: with Computer Excercises*, John Wiley, New York (1968).

APPENDIX: ON THE DEVELOPMENT OF TIMSAC PROGRAM PACKAGES

This appendix is a reproduction of a paper by H. Akaike which was originally published under the same title in Volume 51 of the *Bulletin of the International Statistical Institute*. It is included here to provide a brief review of the application and development of the original TIMSAC program package after the publication of the Japanese version of this book. The authors are grateful to the Institute for granting the permission to include the paper in this book.

Introduction

This paper gives an outline of the series of TIMSAC program packages developed at the Institute of Statistical Mathematics in the past 15 years. Characteristics of each package and earlier examples of application will be briefly described and implications of the experience of the development of these packages on the future development of statistical softwares will also be discussed.

The TIMSAC Packages

The series of TIMSAC (Time Series Analysis and Control) program packages is composed of original TIMSAC, or TIMSAC-71, TIMSAC-74, TIMSAC-78 and TIMSAC-84. The hyphenated numbers denote the year of completion of each package. Each program package is characterized by the use of particular time series models or computational procedures and reflects various phases of the progress of the research on time series at the Institute of Statistical Mathematics. All the programs are written in IBM type Fortran.

The original TIMSAC, or TIMSAC-71, published in Akaike and Nakagawa (1972), contains programs for the spectrum analyses by the conventional windowing procedure and by the autoregressive (AR) model fitting. It also contains programs for the analysis and control of a feedback system by the multivariate autoregressive model fitting.

TIMSAC-74 (Akaike, Arahata and Ozaki, 1975 and 1976) is characterized by the inclusion of procedures for the fitting of autoregressive moving average (ARMA) models. It also contains a program for the analysis of non-stationary time series.

In TIMSAC-78 (Akaike, Kitagawa, Arahata and Tada, 1979) the Householder transformation is systematically used for the fitting of AR models to avoid the computation of autocovariance sequences. Programs for the computation of exact likelihoods of AR and ARMA models are also included.

TIMSAC-84 (Akaike, Ozaki, Ishiguro, Kitagawa, Ogata, Tamura, Katsura, and Tamura 1985) shows a significant departure of the basic modeling from the preceding packages. It contains several programs for the analysis of nonstationary time series by Bayesian type models. Also it contains basic programs for the analysis of point processes and extends the domain of the application of the package from ordinary time series to point processes.

In these packages each program is provided with instructions for its use in the form of comments. TIMSAC-78 and TIMSAC-84 are published with numerical examples to help users in implementing the programs. The list of all the programs in the TIMSAC series is included in Appendix.

Significant Characteristics of the Packages

The most significant general characteristic of the TIMSAC series is the systematic use of the information criterion AIC or ABIC, the Bayesian extension of AIC, for the evaluation and selection of time series and point process models. The model selection in TIMSAC-71 was realized by the use of the final prediction error (FPE) criterion which was later extended to the more general criterion AIC. The model selection is realized by simply choosing a model with minimum FPE or AIC.

TIMSAC-71 is characterized by the systematic use of multivariate AR models for the analysis and control of feedback systems. The AR modeling procedure in the time domain is supplemented by the spectral analysis procedure in the frequency domain that helps the interpretation of the time domain modeling. The optimal controller design procedure through the AR modeling represents a unique contribution of the time series approach to the control problem.

Certainly there is a limitation in the modeling approach due to the choice of the basic family of models. TIMSAC-74 extended the basic model from AR to ARMA. The most serious concern in this case was with the ARMA modeling of vector time series. The problem was the identifiability, or uniquencess, of the representation and a solution was obtained by introducing a state space or Markovian representation (Akaike, 1974).

TIMSAC-74 contains a program, CANOCA, for the preliminary analysis of the structure of the state vector by the canonical correlation analysis of the time series. The program MARKOV fits a Markovian representation with a specified structure of the state vector that is determined by CANOCA or its modification. To the present author's knowledge this was the first procedure of multivariate ARMA model fitting that explicitly took the identifiability problem into account.

The model fitting procedures adopted in TIMSAC-71 and 74 are based on approximate likelihoods that are defined through the sample auto-covariance sequences. TIMSAC-78 bases its procedures mainly on the Householder transformation of the matrix composed of the column vectors defined by successive shifting of the origin of time of the original sequence of observations. By this approach the computation of sample autocovariances is avoided. The approach allows easy adjustment of the lag length of the autoregression for each component of the observation vector.

Although computational accuracy and flexibility of the procedure is increased by this approach the resulting procedure is conditional on the first few observations used to define the initial state of the series. The exact likelihood computation programs for stationary AR and ARMA models included in the package TIMSAC-78 eliminate this problem by introducing the distributions of the initial states explicitly. In particular, the program XSARMA that computes the exact likelihood of a stationary ARMA model contains a subroutine that efficiently computes the covariance matrix of the state vector of a scalar stationary ARMA model by the procedure given in Akaike (1978a).

TIMSAC-78 contains also a program for the analysis of time series with slowly changing spectrum. The program is based on a Bayesian type modeling obtained by defining the likelihood of a model by $\exp(-0.5\text{AIC})$, where AIC is by definition

$$\text{AIC} = (-2) \log \text{ maximum likelihood} + 2 \text{ (number of parameters)},$$

where log denotes a natural logarithm. The use of this type of modeling is discussed in Akaike (1978b, 1979).

TIMSAC-84 is mainly characterized by the use of Bayesian and point process models. The (quasi) Bayesian models provide procedures for the analysis of nonstationarities such as seasonal variations and changing spectrum. The programs for the point process analysis contain those for the trend and cycle analysis of the intensity function of a Poisson process and the analysis of a linearly self-exciting point process with trend and linear input. With other standard programs for the analysis of point processes, such as those for the simulation, graphics and conventional direct Fourier analysis, this collection of point process analysis programs forms a unique contribution that extends the coverage of the application of the TIMSAC series.

The package also contains programs for the fitting of particular nonlinear time series models and a program that is a unification of the analysis and control programs given in the original TIMSAC. One characteristic of this latter program is that it realizes a procedure for the analysis of the behavior of a historical time series data through the simulation of the expected behavior of the time series when the original generating mechanism is modified by the implementation of an optimal controller.

Earlier Examples of Practical Applications

Many of the programs within the TIMSAC program packages were developed by responding to practical needs of applications. In particular, the main part of TIMSAC-71 was developed to provide a procedure for the analysis and control of the cement rotary kiln process. The test with a real process provided an assurance of the success of application of the TIMSAC package in other fields. The package established the use of the multivariate autoregressive model as a basic model for the analysis and control of a feedback system operating under a stochastic environment.

The first example of application was to the analysis and control of a cement rotary kiln process and successful result was reported in Ootomo, Nakagawa and Akaike (1972). The same procedure was applied to the implementation of a ship's autopilot by Ohtsu, Horigome and Kitagawa (1979). This result clearly demonstrated the superiority of this new type of control to that by the conventional autopilot. The procedure was also applied to the control of thermal electric power plants. This produced a first example of practical use of modern optimal control theory in this area and was reported in Nakamura and Akaike (1982).

The technique of the feedback system analysis by TIMSAC-71 package also found important applications. Besides the applications required in the preliminary analyses for the implementation of the above stated controls the technique of the feedback system analysis was applied to the analyses of noisy system as found in experimental atomic power plants (Fukunishi, 1977), economic data (Oritani, 1979), multi-channel record of brain wave (Akaike, 1981), and clinical medical record (Wada, Akaike and Kato, 1986). These applications provided quantitative evidences on the behaviors of the related systems which had hitherto been vaguely guessed by researchers in each subject area.

The program for the analysis of locally stationary process included in

TIMSAC-74 has been adapted to generate a procedure for the automatic detection of the arrival time of an earthquake (Yokota, Zhou, Mizoue and Nakamura, 1981).

The application of the Bayesian seasonal adjustment program BAYSEA, included in TIMSAC-84, to the U. S. economic data revealed limitations of the Census Method X-11 procedure (Akaike and Ishiguro, 1983). Its extension BAYTAP-G for the analysis of earth tide, also included in TIMSAC-84, is now routinely used at the International Latitude Observatory, Mizusawa, Japan, for the analysis of geophysical records related to the deformation of the earth (Ishiguro, 1981).

The programs for the point process analysis in TIMSAC-84 have been extensively applied to the analysis of seismic records by Y. Ogata who developed these programs. One significant application of the program LINLIN is to the analysis of mutual dependence between earthquakes in two different areas of Japan (Ogata, Akaike and Katsura, 1982). The result established the existence of a clear one-sided influence from earth-quakes in the Hida area to those in the Kwanto area. It is expected that the programs will find increasing applications in the earthquake and neural information analyses.

For further publications related to some of the programs in TIMSAC-84 readers are referred to the additional references included in Appendix.

It may be of interest to note here that after 15 years since the introduction of the original TIMSAC an adaptation of the optimal steam temperature control of super critical thermal power plants based on the multivariate AR modeling is now being contemplated in the U. S. for the control of ultra super critical thermal power plants. This provides an example which demonstrates the importance of time series analysis software as a key to the realization of an industrial application of modern optimal control theory.

Implications of the Experience with TIMSAC Packages for the Development of Statistical Softwares

The leading idea for the development of TIMSAC program packages was to produce new models and procedures of time series analysis and provide a proof of feasibility for each procedure in the form of a computer program. The performance of each program has been tested by its application to real or simulated data. Since the main objective was to provide proofs of feasibilities attention was not fully paid to the user friendliness of programs. It was assumed that this aspect would be taken care of by knowledgeable users who would wish to adapt the programs for their own use.

One particular example of adaptation can be seen in the SILTAC (Self-

Instructive, Learning and Turtorial System for Statistical Analysis and Control of Dynamic Systems) system, developed by the System Sogo Kaihatsu Co., Tokyo, which systematically teaches the user of some of the procedures described in TIMSAC in conversational mode on a micro-computer. There remains much to be done in this direction of developing expert systems for time series analysis.

It can fairly be safe to say that any computer program package for time series model fitting cannot be recommended for routine use in time series analysis unless it is equipped with some procedure for the evaluation of the relative goodness of each model. In particular, the difficulty in deciding on the final choice of a model becomes extremely significant with high dimensional vector time series. In our original experience of fitting multivariate AR models to cement kiln process data we had to fit models to the records of 7-dimensional time series of cement kilns under various operating conditions and it became quickly clear that the selection of orders for 7-dimensional AR models was practically impossible without depending on some objectively defined criterion. This lead us to the introduction of the FPE criterion in the original TIMSAC.

Some statisticians consider the model selection procedure by minimizing FPE or AIC as "mechanistic". This represents the lack of experience of handling large number of data sets in a real life situation. The use of artificial intelligence for statistical computing is a popular subject at present. We feel that the model selection by the minimum AIC procedure is an example of successful realization of artificial intelligence that supersedes the activity of ordinary human intelligence in certain situations.

One university professor once jokingly complained about the "automatic" procedure of power spectrum estimation by the minimum AIC procedure of AR model fitting, describing it as a threat to his profession. The complaint meant that before the introduction of the procedure anyone who could properly use the windowing procedure for spectrum estimation was considered to be an expert of time series analysis, while the "mechanistic" procedure eliminated the necessity of such expert. Actually, results obtained by this procedure were often better than those obtained by the windowing procedure in some applications.

During the process of implementing the optimal control of cement kilns results produced by TIMSAC programs clarified some of the limitations of human operators who were the experts of the kiln operation. This suggests possible limitation of the expert system approach to statistical data analysis when it is not supported by any "super intelligent" statistical software.

Statistical procedures are designed to provide proper matching between data from the outside world and human intelligence and aid the development of proper understanding by a human being. Thus a statistical software must be based on a statistical procedure that extends certain aspect of human intelligence to claim its unique existence. Our experience in the development of the TIMSAC packages suggests that the development of proper families of statistical models and criteria for the evaluation of estimated models provides one possibility.

The development of new models will mainly by realized through the contact with real problems and this will also provide a proper framework for the discussion of the choice of model evaluation criteria.

We may thus conclude that the establishment of a proper system that realizes a continuous contact of statisticians with real problems will allow continual development of useful statistical softwares. This was the case with the development of the TIMSAC program packages.

REFERENCES

[1] Akaike, H. (1974) *Markovian representation of stochastic processes and its application to the analysis of autoregressive moving average processes*, Annals of the Institute of Statistical Mathematics, **26**, 363–387.

[2] Akaike, H. (1978a) *Covariance matrix computation of the state variable of a stationary Gaussian process*, Annals of the Institute of Statistical Mathematics, **30B**, 499–504.

[3] Akaike, H. (1978b) *On the likelihood of a time series model*, The Statistician, **27**, 217–235.

[4] Akaike, H. (1979) *A Bayesian extension of the minimum AIC procedure of autoregressive model fitting*, Biometrika, **66**, 237–242.

[5] Akaike, H. (1981) *Recent development of statistical methods for spectrum estimation*, Recent Advances in EEG and EMG Data Processing, Yamaguchi, N. and Fujisawa, K. eds, Elsevier North-Holland Biomedical Press, Amsterdam, 63–78.

[6] Akaike, H., Arahata, E. and Ozaki, T. (1975) *TIMSAC-74, A time series analysis and control program package* (1), Computer Science Monographs, No. 5, The Institute of Statistical mathematics, Tokyo.

[7] Akaike, H., Arahara, E. and Ozaki, T. (1976) *TIMSAC-74, A time series analysis and control program package* (2), Computer Science Monographs, No. 6, The Institute of Statistical Mathematics, Tokyo.

[8] Akaike, H. and Ishiguro, M. (1983) *Comparative study of the X-11 and BAYSEA procedures of seasonal adjustment*, Applied Time series Analysis of Economic Data, Zellner, A., ed. Economic Research Report ER-5, Bureau of the Census, U. S. Department of Commerce, 17–30.

[9] Akaike, H. Kitagawa, G., Arahata, E. and Tada, F. (1979) *TIMSAC-78*, Computer Science Monographs, No. 11, The Institute of Statistical Mathematics, Tokyo.

[10] Akaike, H. and Nakagawa, T. (1972) *Statistical Analysis and Control of Dynamic Systems*, Saiensu-sha, Tokyo. (In Japanese; English translation to be published)

[11] Akaike, H., Ozaki, T., Ishiguro, M., Kitagawa, G., Ogata, Y., Tamura, Y., Katsura, K. and Tamura, Y. (1985) *TIMSAC-84 Part 1 and 2*, Computer Science Monographs, No. 22 and 23, The Institute of Statistical Mathematics, Tokyo.

[12] Fukunishi, K. (1977) *Diagnostic analyses of a nuclear power plant using multivariate autoregressive processes*, Nuclear Science and Engineering, **62**, 215–225.

[13] Ishiguro, M. (1981) *A Bayesian approach to the analysis of the data of crustal movements*, Journal of the Geodesic Society of Japan, **27**, 256–262.
[14] Nakamura, H. and Akaike, H. (1981) *Statistical identification for optimal control of supercritical thermal power plants*, Automatica, **17**, 143–155.
[15] Ogata, Y., Akaike, H. and Katsura, K. (1982) *The application of linear intensity models to the investigation of causal relations between a point process and another point process*, Annals of the Institute of Statistical Mathematics, **34B**, 373–387.
[16] Ohtsu, K., Horigome, M. and Kitagawa, G. (1979) *A New ship's autopilot design through a stochastic model*, Automatica, **15**, 255–268.
[17] Oritani, Y. (1979) *Application of Akaike's method to economic time series*, Kinyu Kenkyu Shiryo, No. 4, Bank of Japan, 103–132.
[18] Otomo, T., Nakagawa, T. and Akaike, H. (1972) *Statistical approach to computer control of cement kilns*, Automatica, **8**, 35–48.
[19] Yokota, T., Zhou, S., Mizoue, M. and Nakamura, I. (1981) *An automatic measurement of arrival time of seismic waves and its application to an on-line processing system*, Bulletin of the Earthquake Research Institute, University of Tokyo, **55**, 449–484. (In Japanese with English abstract)
[20] Wada, T., Akaike, H. and Kato, E. (1986) *Autoregressive models provide stochastic descriptions of homeostatic processes in the body*, Japanese Journal of Nephrology, **28**, 263–268.

Appendix

The programs in the TIMSAC program packages are listed below.

TIMSAC-71

AUTCOR	(Autocovariance function computation (uni-variate); direct method)
MULCOR	(Crosscovariance function computation (multi-variate); direct method)
FFTCOR	(Crosscovariance function computation (bi-variate); FFT method)
AUSPEC	(Power spectrum computation; uni-variate)
MULSPE	(Cross spectrum computation; multi-variate)
SGLFRE	(Frequency response function computation; single-input)
MULFRE	(Frequency response function computation; multi-input)
FPEAUT	(FPE Computation for uni-variate autoregressive model)
FPEC	(FPEC Computation for control system model or multi-variate autoregressive model)
MULNOS	(Relative power contribution computation)
DECONV	(Inpulse response computation)
RASPEC	(Rational spectrum computation; uni-variate)

MULRSP (Rational spectrum computation; multi-variate)
OPTSIM (Optimal control simulation)
WNOISE (White noise simulation)

TIMSAC-74

PART-1

CANARM (Canonical correlation analysis of scalar time series)
AUTARM (Automatic AR-MA model fitting; scalar case)
COVGEN (Covariance generation from gain function)
CANOCA (Canonical correlation analysis of vector time series)
MARKOV (Maximum likelihood computation of Markovian model)

PART-2

PRDCTR (Prediction by AR-MA model)
SIMCON (Optimum controller design and simulation)
NONST   (Non-stationary power spectrum analysis)
PWDPLY (Power spectrum display)
FRDPLY (Frequency response function display)
THIRMO (Third order moment computation)
BISPEC (Bi-spectrum computation)

TIMSAC-78

UNIMAR (Univariate case of minimum AIC method of AR model
        fitting)
UNIBAR (Univariate case of Bayesian method of AR model fitting)
BSUBST (Bayesian type all subset analysis of time series by a model
        linear in parameters)
MULMAR (Multivariate case of minimum AIC method of AR model
        fitting)
MULBAR (Multivariate case of Bayesian method of AR model
        fitting)
PERARS (Periodic autoregression for a scalar time series)
MLOCAR (Minimum AIC method of locally stationary AR model
        fitting; scalar case)
BLOCAR (Bayesian method of locally stationary AR model fitting;
        scalar case)

MLOMAR (Minimum AIC method of locally stationary multivariate AR model fitting)

BLOMAR (Bayesian method of locally stationary multivariate AR model fitting)

NADCON (Noise adaptive controller)

EXSAR (Exact maximum likelihood method of scalar AR model fitting)

XSARMA (Exact maximum likelihood method of scalar AR-MA model fitting)

TIMSAC-84

PART 1

BAYSEA (Bayesian seasonal adjustment)

BAYTAP-G (Bayesian tidal data analysis)

DECOMP (Time series decomposition into components by the Bayesian approach)

LOCCAR (Locally constant AR model)

TVCAR (Time varying coefficient AR model)

NONSPA (Nonstationary spectrum analysis by minimum ABIC procedure)

PART-2

MULCON (Multiple time series analysis by simulated prediction and control)

SNDE (Stochastic nonlinear differential equation model)

ADAR (Amplitude dependent AR model)

EPTREN (Exponential polynomial or Fourier series modeling of trend and cycle of Poisson intensity)

LINLIN (Linearly self-exciting process with trend and linear input)

PGRAPH (Graphic point process analysis)

LINSIM (Simulation of the point process identified by LINLIN)

SIMBVH (Simulation of a bi-variate Hawkes point process)

PTSPEC (Point process spectrum by direct Fourier transform)

PUBLICATIONS RELATED TO TIMSAC-84

BAYTAP-G

Ishiguro, M. Akaike, H., Ooe, M. and Nakai, S. (1983) A Bayesian approach to the analysis of earth tides, Proceedings of the 9-th International Symposium on Earth Tides, Kuo, J. T., ed., E. Schweizerbart, Stuttgart, Germany, 283–292.

DECOMP

Kitagawa, G. (1981) A nonstationary time series model and its fitting by a recursive filter, Journal of Time Series Analysis, 2, 103–116.

Kitagawa, G. and Gersch, W. (1984) A smoothness priors - state space modeling of time series with trend and seasonality Journal of the American Statistical Association, 79, 378–389.

TVCAR

Gersch, W. and Kitagawa, G. (1985) A smoothness priors time-varying AR coefficient modeling of nonstationary covariance time series, IEEE Transaction on Automatic Control, 30, 48–56.

Kitagawa, G. (1983) Changing spectrum estimation, Journal of Sound and Vibration, 89, 433–455.

SNDE

Ozaki, T. (1985) Nonlinear time series models and dynamical systems, Handbook of statistics, Vol. 5, Hannan, E. J., Krishnaiah, P. R. and Rao, M. M., eds., North-Holland, Amsterdam, 25–83.

ADAR

Haggan, V. and Ozaki, T. (1981) Modelling nonlinear random vibrations using an amplitude-dependent autoregressive time series model, Biometrika, 68, 189–195.

LINLIN

Ogata, Y. (1983) Likelihood analysis of point processes and its applications to seismological data, Bulletin of the International Statistical Institute, 50, Book 2, 943–961.

Ogata, Y. and Katsura, K. (1986) Point process model with linearly parameterized intensity for the application to earthquake data, Essays in Time Series and Allied Processes, Gani, J. and Priestley, M. B., eds, Journal of Applied Probability 23A, 291–310.

Matsumura, K. (1986) On regional characteristics of seasonal variation of shallow earthquake activities in the world, Bulletin of the Disaster Prevention Research Institute, Kyoto University, 36, 43–98.

PGRAPH

Ogata, Y. and Shimazaki, K. (1984) Transition from aftershock to normal activity: The 1965 Rat Bulletin of the Seismological Society of America, 74, 1757–1765.

Matsumura, R. S. (1986) Precursory qiescence and recovery of aftershock activities before some large aftershocks, Bulletin of the Earthquake Research Institute, University of Tokyo, 61, 1–65.

LINSIM

Ogata, Y. (1981) On Lewis's simulation method for point processes, IEEE Transaction on Information Theory, IT-27, 23–31.

Summary

The program packages of the TIMSAC series have been developped at the Institute of Statistical Mathematics since 1971. In this paper the construction of the packages is explained and early examples of their applications are presented. The paper concludes with the discussion of implications of the experience of the development of these packages on the future development of statistical softwares in general.

Résumé

Les logiciels des séries TIMSAC ont été développés à l'Institut des Mathématiques Statistiques depuis 1971. Ce rapport expose la construction des logiciels et présente les premiers exemples des leurs applications. La conclusion traite des implications de l'expérience du développement de ces logiciels sur le développement futur des logiciels statistiques en général.

INDEX